About Island Press

Since 1984, the nonprofit organization Island Press has been stimulating, shaping, and communicating ideas that are essential for solving environmental problems worldwide. With more than 1,000 titles in print and some 30 new releases each year, we are the nation's leading publisher on environmental issues. We identify innovative thinkers and emerging trends in the environmental field. We work with world-renowned experts and authors to develop cross-disciplinary solutions to environmental challenges.

Island Press designs and executes educational campaigns, in conjunction with our authors, to communicate their critical messages in print, in person, and online using the latest technologies, innovative programs, and the media. Our goal is to reach targeted audiences—scientists, policy makers, environmental advocates, urban planners, the media, and concerned citizens—with information that can be used to create the framework for long-term ecological health and human well-being.

Island Press gratefully acknowledges major support from The Bobolink Foundation, Caldera Foundation, The Curtis and Edith Munson Foundation, The Forrest C. and Frances H. Lattner Foundation, The JPB Foundation, The Kresge Foundation, The Summit Charitable Foundation, Inc., and many other generous organizations and individuals.

The opinions expressed in this book are those of the author(s) and do not necessarily reflect the views of our supporters.

A Good Drink

A Good Drink

In Pursuit of Sustainable Spirits

Shanna Farrell

ISLANDPRESS | Washington | Covelo

Library of Congress Control Number: 2021934385

All Island Press books are printed on environmentally responsible materials.

Manufactured in the United States of America
10 9 8 7 6 5 4 3 2 1

Keywords: Bar Agricole, bartending, brewery, cocktail culture, craft distillery, Equiano, farm bar, farm-to-table, fermentation, food system, food waste, good food movement, High Wire Distilling Co., Jimmy Red corn, Leopold Bros., liqueur, local sourcing, Maker's Mark, malting facility, mescal, mezcal, Mezonte, mono-cropping, Montanya, Mr Lyan group, responsible sourcing, St. George Spirits, sustainable agriculture, Tales of the Cocktail, terroir, Tin Roof Drinks, Trash Collective, whiskey

Once the World Was Perfect

Once the world was perfect, and we were happy in that world.
Then we took it for granted.
Discontent began a small rumble in the earthly mind.
Then Doubt pushed through with its spiked head.
And once Doubt ruptured the web,
All manner of demon thoughts
Jumped through—
We destroyed the world we had been given
For inspiration, for life—
Each stone of jealousy, each stone
Of fear, greed, envy, and hatred, put out the light.
No one was without a stone in his or her hand.
There we were,
Right back where we had started.
We were bumping into each other
In the dark.
And now we had no place to live, since we didn't know
How to live with each other.
Then one of the stumbling ones took pity on another
And shared a blanket.
A spark of kindness made a light.
The light made an opening in the darkness.
Everyone worked together to make a ladder.
A Wind Clan person climbed out first into the next world,
And then the other clans, the children of those clans, their children,
And their children, all the way through time—
To now, into this morning light to you.

Contents

Conclusion

In Pursuit of Sustainable Spirits

In my early twenties, during the nights I spent in dark, sweaty rock clubs where the walls pulsed in time with the bass guitar, I exclusively drank Jack and Gingers. As I got a little older and traded bars with loud music for those with flickering candles and leather furniture, I upgraded to Whiskey Sours. When I got behind the bar myself, shaking tins and gliding long spoons through liquid and ice, I switched to Old Fashioneds. It became part of my job to taste spirits, learn their stories, and sell them to guests.

The stories that stuck with me always concerned how the spirits were made. I often forgot the proof or price of a particular brand, but I'd remember minute details about the water used to distill or bottle it. I preferred the spirits whose labels told us about the hands that created them and the ingredients from which they were born.

Many years ago, one whiskey in particular caught my eye. It had been on the shelf since the little corner cocktail bar where I worked had opened, but it was expensive and my humble graduate school bank account didn't allow me to buy a pour. I didn't try it until months after I started working there, instead settling for holding the elegant bottle in my hands and reading the label over and over. It was made with 100 percent New York State grains, something I'd never seen advertised before. When I finally did try it, the flavor was hot and unbalanced, but the grain fought its way through, adding depth to the flavor. Though it didn't turn out to be my favorite whiskey, it started me thinking about the crops that were used to make it.

Later, in my early thirties and now an interviewer with UC Berkeley's Oral History Center, I became curious about cocktail culture and the sustainability of the spirits industry. I interviewed distillers, bartenders, and cocktail historians. I learned the details of distilling and how the grains used to make alcohol are grown. I read about the effects of mono-cropping on farmland; considered the tremendous amount of wastewater created by distilleries; and saw the mountains of trash thrown out by bars each night. I watched as some brands began to tout their eco-consciousness. I noticed as trends came and went. I lived through seasons of drought, rain, and wildfire. I wondered what the disrupted climate meant for the future of alcohol.

In 2016, I found myself in Charleston, South Carolina, at a beverage conference aptly called BevCon. I was among others who, like me, were curious about issues facing the spirit industry. Though I had friends and colleagues in attendance, one specific seminar caused me to fly across the country in August to a place where the air was thick enough to cut with a knife. The session was called "Drinking as an Agricultural Act," and, on the last day of BevCon, I sat in a hotel conference room with the air-conditioner soothing my sunburnt skin, listening to a beer brewer, a cider maker, and a whiskey producer talk about the crops they used to make their drinks.

Ann Marshall, co-owner and distiller of High Wire Distilling Co., described planting an heirloom corn that had teetered on the brink of extinction, working with farmers to manage the crop, and gathering a group of friends to harvest it before a hurricane blew in. To her, whiskey was an agricultural product and, therefore, it was tied to environmental health. Just as we need biodiversity in wild plants, we also need it in cultivated crops. Among other benefits, genetic diversity in the foods we eat and the drinks we sip protects against crop diseases, improves soil health, and creates resilience to climate change.

Yet almost all corn-based whiskey is sourced from a single variety: yellow dent field corn. Ann and her husband, Scott Blackwell, with whom she owns High Wire, didn't want to use it. Instead, they opted to plant Jimmy Red corn, a legendary moonshiner's corn that had dwindled down to two cobs following the death of the last man known to grow it. Starting with just two and a half acres, Ann and Scott set out to make a small dent in yellow dent's stranglehold on the whiskey market.

With her Southern drawl and her excitement about Jimmy Red corn, Ann sparked something in me that day. I knew I needed to try her bourbon to see if it lived up to its story. When I finally got my hands on some, the first thing I did was read the bottle, letting the anticipation build. It read: "NEW SOUTHERN REVIVAL / STRAIGHT BOURBON WHISKEY / MADE WITH 100% JIMMY RED CORN / DISTILLED WITH PRIDE BY HIGH WIRE DISTILLING CO." I turned the bottle over to read the other side. "New Southern Revival Brand is a celebration of the diverse agricultural traditions of our region. A true revival spirit, this whiskey began as a labor of love to save Jimmy Red corn from near extinction."

The more I read, the more I liked it. Learning its history made me feel connected to this whiskey. I swirled the liquid around in my glass and immediately noticed the viscosity. It was slightly thicker than average, which made it almost slick on my tongue, like glycerin but with none of the soapy taste. It reminded me of nuts, banana, and grass, with a hint of brine and minerality. At 102 proof, I expected it to be hot, but instead it slid down my throat with no trace of burn. I had never tasted anything like it. It felt special. I felt special for drinking it.

That feeling of connection is at the heart of *terroir*, the characteristic taste and flavor imparted to a crop by the environment where it was grown, linking a spirit to the place where it was created. Maybe the search for connection is why alcohol aficionados tend not just to enjoy a good drink, but also to want to know everything about it. Indeed, a whole genre of writing has taught us about the history of spirits, the science behind how they are made, how they should taste, and what cocktails they complement best. But such curiosity hasn't usually extended to how alcohol fits into our food system, and what that means for the environment.

As eaters, we have become aware of the environmental problems created by industrial agriculture, and we have started to question where our food comes from. We are concerned about the health effects of pesticides, the dead zones created by chemical fertilizers, and the carbon footprint of shipping out-of-season food all over the planet. Each of these problems dogs the spirits industry, just as it does the food system. But while we may insist on organic, locally grown produce, we've yet to engage with spirits at the same level. We know some distillers, but not the farmers who supply them with the ingredients that enable their work.

Part of the issue may be that we simply don't see spirits as food, though they come from the same crops as those that feed us at our dinner tables. They aren't regulated the same way, by the same governmental department, with the Food and Drug Administration being responsible for oversight of food production, and the Alcohol and Tobacco Tax and Trade Bureau for oversight of spirits. And not everyone drinks alcohol, which we don't need for survival like we do food. These factors drive a wedge between how we treat food and how we treat alcohol.

This disconnect extends to restaurants, too. In the age of celebrity chefs and farm-to-table eating, restaurants are lauded for the way they source their ingredients. These days, most of the menus I'm handed proudly declare the restaurant's purveyors; restauranteurs know that customers appreciate transparency. I now instantly recognize the names of farms. I can visit their websites, read about who works there, and look at pictures illustrating their operations. It makes me feel good about what I'm eating—that I'm being responsible, that I'm connected to them in some way.

But this doesn't happen at the bar. My years of bartending have taught me that the big brands found in nearly every cocktail bar across the globe—Campari, Aperol, Luxardo Maraschino—are mass-produced and full of artificial ingredients. Their production methods contribute to climate change, water pollution, and pesticide resistance. This wouldn't be tolerated in the farms that supply the ingredients used in the kitchen. Yet at the bar—the profit center of a restaurant—sustainability is an afterthought. Nine times out of ten, I'm disappointed when I scan the bottles at a high-end bar or read the cocktail menu at a restaurant known for their responsible sourcing.

After my trip to Charleston, I was left with questions: Can spirits be produced without harming the environment? Who are the eco-conscious producers? Why does sustainability matter to them? What does it mean to be truly sustainable? How can we, as consumers, support their efforts? What will it take for us to think about alcohol the same way we think about food?

It's these questions that led me to different parts of the world, to conversations with people who are thinking deeply about these issues. They led me to Charleston, where it all started, to visit a farm where Jimmy Red corn grows. They took me to Guadalajara, Mexico, where I met distillers and farmers working to preserve traditional methods of making mezcal and sustainable ways of growing agave. I flew to Denver, Colorado, to visit the only distillery in the United States that malts its own barley and uses solar power. I drove across the Bay Bridge to Alameda, California, to talk with brandy producers about how climate change affects their supply chain. I went to Portland, Oregon, to learn about training programs that teach bartenders around the world to reduce waste. I had coffee with a bartender who works at a "farm bar" that carries only spirits made with responsibly grown ingredients. I went to Kentucky to visit Maker's Mark and see if it's possible to preserve local ecology while producing spirits on a large scale.

And then the global COVID-19 pandemic hit. This left us all grounded and the industry reeling, trying to make sense of the present and how to move forward in the future. My travels became virtual. I spoke to a rum maker to learn about the spirit's fraught colonial history and his company's decision not to use additives. I interviewed a distiller about sourcing American-grown sugarcane and the process of becoming a certified B Corp. I talked to one of the most acclaimed bar operators in the business, who has made a career of combining sustainability and luxury.

In short, I went in search of a good drink. And I found it in abundance. I met people who are sourcing, distilling, and bartending sustainably, creating models that will redefine the industry. They, along with many others who aren't featured here, are transforming how spirits are made—and the consequences for people and the planet—one bottle at a time.

CHAPTER 1

Whiskey

"I brought bourbon to the marriage," Ann Marshall tells me from across the table where she is seated next to her husband, Scott Blackwell. It's 2017, roughly a year after I heard Ann speak at the BevCon meeting about using local organic grains to make whiskey. We are on the second floor of their distillery, a former Studebaker car dealership in Charleston, South Carolina, overlooking a chiseled fermentation tank, a tall copper still, a short bottling line, and tiered rows of barrels where their bourbon is aging.

When she was a child, Ann's father would come home at the end of each day and make an Old Fashioned for himself and one for her mother. They'd steal off into the study, cocktails in hand, and talk for an hour before dinner. It was a family tradition that Ann remembers as the most peaceful time of day, one that marked the beginning of her relationship with bourbon. As she got older, she adopted this as her own evening ritual. "There is something so settling about coming home and having a dram of whiskey," she says. When she and Scott were first dating, she would invite him over for a pre-dinner drink. It was these moments that turned him into a card-carrying whiskey enthusiast and that built the foundation of High Wire Distilling Co.

Before they had whiskey, the couple had cookies. They met at Immaculate Baking, which holds the Guinness World Record for making the largest chocolate-chip cookie in the world. The bakery was the latest of Scott's business ventures; after selling pies out of his garage in college, he started several restaurants and a coffee roastery. Straight out of Duke University, Ann began working as the head of marketing before moving on to another natural-foods company. Immaculate Baking did so well that it caught the eye of General Mills, and selling the bakery gave Ann and Scott the funds to break into the spirits industry. In 2013, they opened High Wire in a gentrifying section of downtown Charleston, a few blocks from the bars and restaurants that now identify as Upper King Street.

Ann and Scott are part of a Southern farm-to-table culinary scene that promotes local, seasonal, and traditional growing methods and includes not only chefs, but also artisanal producers, historians, brewers, seed collectors, and farmers. They often say that they are food people first, and their network of friends reflects this assertion.

Yet despite this community, Ann and Scott didn't initially see the connection between distilling and the good food movement. Sustainability wasn't their focus, and they didn't realize that they could source their raw ingredients from nearby farmers.

"My family has been in the farming business for centuries, and all they grow now are row crops, things you can't eat, like ethanol corn or soybeans for oil and cotton," Ann says. "We just weren't sure if we were actually going to use local grains. We didn't know that they were available or that we could orchestrate a contract farming situation. It was not really something we felt was very possible."

While Ann knew that many chefs and restaurants work closely with farms and are able to order specific products from them (contracting with them for produce or protein), this isn't common in the spirits industry. Most distillers buy their grain in bulk from distributors who are effectively intermediaries, leaving Ann no reason to think that sourcing could happen another way.

Those distributors buy from commodity farmers, who live and die by supply and demand. To keep up a steady supply of their crop, the farmers grow monocultures of a single plant, such as the corn used to make whiskey. Mono-cropping, currently practiced on more than half of all US farmland, strips the soil of nutrients, leaves crops vulnerable to disease, and typically relies on pesticides and chemical fertilizers. It's part of a broken industrial system that damages the land and keeps farmers beholden to giant agribusinesses, relying on government subsidies in order to stay afloat. Farmers must plant, plant, plant and grow, grow, grow in a never-ending race to keep ahead of the weeds and creditors.

Distillers are just beginning to recognize their role in that system. In late 2020, the Distilled Spirits Council of the United States formed a new environmental sustainability working group to share best practices on issues including land stewardship, responsible water use, and waste reduction. The president of the industry group, Chris Swonger, spoke in tones similar to Ann's: "Distilled spirits are agriculturally based products. . . . Efficiency matters, and from field to bottle, every drop counts."

But in 2013, when Ann and Scott were getting High Wire off the ground, sustainability was a lonely row to hoe.

～

There are a few things to know about whiskey. First, *whiskey* is an umbrella term for a few types of liquors, including bourbon, scotch, and single malt. Next, it has to be made from grain, such as corn, barley, rye, and wheat. Lastly, it's usually aged in wooden barrels.

Bourbon has its own set of rules that go beyond those few basic qualifications (as does scotch, which has to be made in Scotland, otherwise it's called a single malt). Bourbon, which is produced only in the United States, has to be made with mostly corn—that is, at least 51 percent of its mash bill, or recipe. The other 49 percent can be any other type of grain, but the mix usually includes barley to help with the fermentation process, which turns sugar into alcohol. Bourbon has to be aged in new charred-oak barrels. Straight bourbon—a subcategory of the spirit—must be aged for at least two years and no more than four.

The list continues. When bourbon is distilled, it can't be higher than 160 proof, or 80 percent alcohol by volume (commonly referred to as ABV, which is always half the proof). When it goes into a barrel to be aged, water must be added, reducing the proof to 125, or 62.5 percent ABV. When it's bottled, it must be at least 80 proof, or 40 percent ABV. The bottle must have an age statement on it, indicating how much time it spent inside an oak barrel. If a spirit doesn't meet these requirements, it's just called whiskey (and may be a blended whiskey or a single malt, each of which has another set of rules).

Bourbon's roots are in the South—Kentucky, to be exact. While the precise origin of bourbon is up for debate, it's believed that an early form of the whiskey was brought over to America by the Scots. There are several bourbon companies that claim to have invented it, but without proof this remains a myth. Though its origin is enigmatic, a few things are certain: it became popular during the American Revolution; corn grows well in Kentucky; and the spirit officially began to be called "bourbon" in the 1850s. And, like many things, much of it was produced by slave labor before the Civil War and by formerly enslaved people after the war ended.

Though the connection between distilling and slave labor hasn't been discussed much until recently, tax records from the eighteenth and nineteenth centuries reveal that distillers claimed enslaved people as property. Auction rolls reveal that white men claimed enslaved people's knowledge of making alcohol as an asset, with ads marketing slaves as "skilled whiskey distillers." Just as enslaved people were forced to work in fields, building the foundation of our food system, and in kitchens, cooking food for white families, they were also put to work making whiskey and rum, the two most popular spirits of the time.

But this aspect of history, like many others, was largely swept under the rug, with many white distillers preferring to take credit for this work, and in the twentieth century, to wipe this history from their company's collective memory.

Distilleries are only now making their role in the slave economy public, trying to control the narrative with over 150 years of distance. In 2016, Jack Daniel's opened up about how their founder, Jack himself, had learned to make whiskey from Nathan "Nearest" Green, a man enslaved by local preacher Dan Call, who, since 1866, had been credited by the company as the person who taught Jack to distill. This confession came in the form of a 2016 *New York Times* article written by Clay Risen, who asserted: "Enslaved men not only made up the bulk of the distilling labor force, but they often played crucial skilled roles in the whiskey-making process."

Since then, companies including Jacob Spears, Elijah Craig, and Henry McKenna have come clean about using slave labor in their early days. University professors such as Wiggins Gilliam have begun to research historic documents and photographs, dusty after years of being buried in archives, of enslaved Black men who helped shape the whiskey industry. A 2019 exhibition in the Frazier History Museum in Louisville featured the stories of enslaved people who were forced to work in distilleries. That same year, the Kentucky Black Bourbon Guild was founded in Lexington. And many of us might have heard that George Washington used slave labor at his Mount Vernon distillery, a history that is becoming common knowledge.

Over the years, as bourbon grew in popularity and more distilleries opened, the industry needed to keep up with demand. Many companies scaled up. This meant that they started making bourbon in larger and larger batches, and with an eye toward their bottom line, using cheaper ingredients. Enter yellow dent, a type of corn that is uniform in size and has a high soft starch content, making the sugars perfect for fermentation. Its flavor is quite consistent, which works well for industrial uses like animal feed, ethanol, high-fructose corn syrup, cooking oil—and bourbon.

Yellow dent corn gets its name from a small indentation that marks the crown of each kernel. It's a hybrid corn that was created in 1846. James L. Reid, a farmer who made his way to Illinois from Ohio, crossed Johnny Hopkins—a red corn—with flint corn and floury corn. The result was yellow dent. The new variety became popular with farmers because it was easy to grow, and it won a prize at the 1893 World's Fair. Today, most hybrid corns and cultivars are offshoots of it, including popcorn, flour corn, and sweet corn.

Much of the industry began using yellow dent after World War II, when many farms started adopting industrial practices. Yellow dent, a commodity crop, is inexpensive to grow and can be produced in mass quantities. It is bought, sold, and traded on the open market, its value fluctuating with supply and demand. It's usually farmed with the use of pesticides, these chemicals blanketing fields and seeping into groundwater. Yellow dent is also frequently overplanted, causing the soil to lose nutrients, and farmers seldom rotate it with cover crops, so fields can't recover their health.

There is never a shortage of yellow dent. Distillers never have to worry about their supply of corn. Because it's cheap and always available, it's basically ubiquitous. As a result, many bourbons have a similar, consistent flavor. Distillers try to distinguish the taste of their products through the little details they can control, such as the yeast strands they use, how long they age their bourbon, and the char levels on the oak barrels—but not the quality of the raw ingredients.

And because so many whiskey producers started using yellow corn, many of them lost connection to the farms where their crops are grown. More often than not, they don't even know which state the grain comes from. A few years ago, I went to Kentucky and Tennessee to travel the Bourbon Trail, bouncing from one distillery to the next over the course of five days. I'd often ask the tour guide where they got their grain and invariably the answer would be "the Midwest." Sometimes, if I was feeling bold, I'd keep asking about it.

"Do you know where in the Midwest?" I'd ask.

"Either Illinois or Indiana" was their response.

Or "I'm not sure."

Or "I'd have to ask the distiller."

Or "One of the square states."

Perhaps this is why most distillers don't focus on ingredients. Perhaps this is why they'd rather talk about yeast, or aging, or char. Perhaps this is why we don't treat alcohol as an agricultural product.

We lose a lot when we forget that spirits are food. We don't see the people or places that shape the American spirits industry. We don't understand the consequences of production. We don't realize that chemicals many of us don't tolerate in our foods—like red dye no. 40

and high-fructose corn syrup—end up in our cocktails. We don't know which products to vote for with our consumer dollars. We're not seeing whiskey for what it is, and what it could be.

~

While it took some time for Ann and Scott to realize their role in the good food movement, their culinary past was never far behind them. And certainly not in South Carolina, where heirloom foods were having a moment and consumers were beginning to question industrial farming. John T. Edge, executive director of the Southern Foodways Alliance, part of the Center for the Study of Southern Culture, and a longtime friend of Ann and Scott, told them about a farm in Tennessee that was growing sorghum, a drought-resistant cereal grain known for its sweetness.

The farm, Muddy Pond, is run by a Mennonite family. They do much of the work to grow and harvest their sorghum by hand, including cutting the cane. They don't use any additives for color or flavor. They mill and press the cane at the farm, where they bottle the juice. In every way, their process is a rejection of industrial farming. At John T.'s urging, Ann and Scott gave it a try, thinking it might make for an interesting whiskey.

"We ordered a fifty-gallon drum of sorghum syrup, which is just the richest, most phenomenal syrup you've ever eaten. It's replaced all cane-based syrup in our house," Ann tells me. "We distilled it and it smelled like apples. It had this beautiful nose."

"That was really when the lightbulb went off," Scott adds. "Like when you put this ingredient in and you leave it alone, just add yeast and do as little to it as possible, and when it comes off the still it tastes completely different than other ingredients. It's not just alcohol."

"That was really the biggest 'aha' moment," Ann says. "The taste is of that farm. Like, what's around it? Is there an apple orchard next to it? And what's contributing to this? They have a very specialized process. It's a Mennonite farm. They plow the fields with draft horses. They do most of their harvesting with a tractor. They incorporate the grain in the syrup, so it's the whole plant."

Ann describes the flavor as malty and nutty and sweet. It's a symphony of flavors that creates complexity, especially when the whole plant is used, which is what technically qualifies it as whiskey. "It's got these beautiful protein strands in the syrup that translate straight through the distillate."

This experience changed everything. It was the first time the couple had used a single farm for ingredients, and the whiskey's terroir could be traced to an exact location. They appreciated the environmentally friendly approach that the family took to growing sorghum, even if they did it more for tradition than for the planet. Their sorghum whiskey is now one of High Wire's staples.

On the heels of the sorghum whiskey came the watermelon brandy, distilled from the Bradford melon, one of the three oldest surviving watermelon varieties in North America. Like many plants and spices, this melon has its origins in the West Indies, its predecessor traveling by boat—a British prison ship, to be exact—during the final months of the Revolutionary War in 1783. The seeds were collected, saved, and planted in Georgia by an American soldier who had been captured by the British. In about 1840, a South Carolinian named Nathaniel Napoleon Bradford crossed that melon with another, the Mountain Sweet. With that the Bradford watermelon was born.

The Bradford watermelon became known for its sweet, fragrant flesh. Its level of sweetness rates a 12.5 on the Brix scale, which measures soluble solids in liquid. Most melons rate a 10. The Bradford's juice was used for molasses, made into syrup, and distilled into brandy. By 1860, it was the most popular late-season melon in the South. But by 1922, the last commercial crop was planted as farmers began to prefer melons with a harder exterior. After that, Nathaniel Bradford's descendants saved the seeds and planted them in their backyards.

Enter Dr. David Shields, a historian, professor at the University of South Carolina, and writer who authored *Southern Provisions: The Creation and Revival of a Cuisine* (2015). In 2003, he began working with Glenn Roberts, founder of Anson Mills, a company dedicated to preserving food heritage through native seed collecting. Glenn's interest in antebellum heirloom crops began with rice and, later, corn. He would drive down rural roads looking for pre–Civil War corn varietals. In 1997, this search led him to a bootlegger's field in South Carolina, where he learned about the potential for commercial distilling of landrace grains—ancient pre-hybridized varietals, such as Jimmy Red corn, with distinctive characteristics that have developed over time as plants have adapted to the conditions of a localized region. Together, the two men studied the history of these grains and how they might be preserved. Dr. Shields took an academic approach while Glenn was the boots on the ground, enlisting farmers who would help him grow landrace grains.

In 2005, Dr. Shields became interested in the Bradford watermelon. He began to ask around to see if anyone was holding onto seeds of the melon, then thought to be extinct. Nothing came of his inquiries until 2012, when he tracked down Nat Bradford, the great-grandson of Nathaniel Bradford. Nat, a young man in his thirties, had saved the seeds, which have been passed down from generation to generation. With a nudge from Dr. Shields, Nat planted some of the melon seeds in 2013. That summer, the seeds grew into 440 plants, yielding 465 melons. Just like that, they were bringing the Bradford watermelon back from the brink of extinction.

Nat brought the melons to Charleston as soon as they were harvested. He gave 50 of them to chef Sean Brock, who turned them into molasses and pickles for McCrady's, one of his restaurants. He gave 140 melons to Ann and Scott.

The couple spent twelve hours cutting and mashing the sweet, fleshy melons. They distilled them into 143 bottles of brandy, a limited edition. "It's like eating a watermelon in reverse," Scott says of its flavor. (It's true—I've tried it and can't think of a better way to describe the way it tastes.) They sold each bottle for $79 until they ran out.

After that, the couple never looked back. Working with small farms and heirloom varietals would be their future.

~

When I found myself back in Charleston, my mission was singular: to find out if High Wire's Jimmy Red Corn Bourbon was indeed superior, both in terms of flavor and how it was produced. Was it everything that it was made out to be?

I learned that, although the story of Jimmy Red began long before Glenn Roberts had come into the picture, there's not a whole lot known about it. Here's what we do know: at some point, it made its way from Georgia to James Island, off the coast of South Carolina, its place of origin unknown. It was used for moonshine—liquor that was produced illegally—prior to the Civil War, and some families kept this tradition alive as the nineteenth century stretched into the twentieth. It was just called "that red corn" among locals looking for a good moonshine variety, referred to as "scrap iron."

As the decades wore on, Jimmy Red dwindled with moonshining. The corn was down to the last two known cobs when a deceased moonshiner's surviving children became concerned about its future. Their father had been making whiskey from it his whole life in the family's backyard, and they didn't want it to die out. They contacted Ted Chewning, a James Island native, and gave him the last two cobbs. Another game of telephone tag ensued. Ted, who knew Glenn Roberts, gave him some of the seeds. Glenn, knowing that chefs love rare ingredients, contacted chef Sean Brock. They started to work with two academics, David Shields and Brian Ward from Clemson University, who were able to grow it on their research farm. Once there was enough to work with, Brock used it for cornmeal in dishes he served at his restaurants, including Husk and McCrady's.

"Sean really kickstarted the Jimmy Red phenomenon and got it into the Charleston lexicon by putting it on the plate," Ann says.

This was all Ann and Scott knew when Glenn told them it would make a great corn for bourbon. They were intrigued and excited, but not enough of it existed for them to turn Jimmy Red corn into whiskey.

"We needed a thousand pounds, and it hadn't yet been grown to that scale," Ann says. They'd need to will it into the world if they wanted to see how it tasted in liquid form. Which is exactly what they did.

Beginning in 2014, the couple worked with Dr. Shields to research the history of the corn while Glenn and Brian helped grow the corn. They used two and a half acres of Clemson's research farm to grow a crop of Jimmy Red corn. Ted Chewning also appeared from time to time to check in on their progress. Their relationship to the crop grew from experimental to intimate. Scott and Ann were at the farm constantly, driving the fifteen miles from Charleston each day to help cultivate the seed at every stage of the corn's life cycle.

"I went out to the field and ran behind the tractor as the seeds were put into the ground, watching it germ, and grow up to my knee. Ann and I were out there while they were flaming the fields at 5:30 in the morning," Scott remembers.

Ann compares it to a first baby. "We have pictures of every single move it made," she says. The corn grew and grew, the stalks rising high into the flat field.

And then the rain came. It poured. It didn't stop. Just a week before harvest, it kept the team from getting into the field. They waited. And waited. Once the sky finally cleared, leaving the earth saturated and the roads slick, Brian called them.

"Do you guys have any friends?" he asked. "Because I think we're going to need to get this out by hand."

Person after person descended onto the farm while the sun did its best to dry out the wet husks. Everyone got their own row. They picked and they picked as the sound of their voices carried through the quiet field. By the end of the day, they had cleared all two and a half acres.

"It was so special for that to be our first harvest," Ann says. "I really knew what community farming was all about."

A change comes over the couple as they tell this story while surveying the distillery they had built. Their speech quickens, their gestures become animated, and light fills their eyes. It's this part of the distilling process that excites them. It's the planting of seeds, tending to the little green sprouts that transformed into tall brown stalks, running after the tractor in the dark, and gathering a group of friends to help free the cobs from their husks. This part—the farming—might seem like the most daunting aspect of making spirits, and one we don't often associate with distilling. But, for Ann and Scott, it was the most rewarding. The raw material, the food, gave them license to pour themselves into the whiskey.

They were doing their part to preserve history, blemishes and all, so that future generations would be able to enjoy this corn on land preserved from the ravages of machines or chemicals. They did the labor and developed the skills that farming requires, saw why sustainable practices are so important, and weighed the trade-offs that come from scaling up and growing an easier, more reliable crop.

Many distillers—especially those who are just getting started—default to the industry standard, yellow dent corn. It grows reliably and in abundance. It's easy to manipulate, much of it genetically modified. Its taste doesn't vary. There's no mystery, no romance. It's safe, which is why Ann and Scott started out with it themselves. But later, when they realized they could do more with heirloom seed varietals, they decided to risk months in the field with their days full of blood and sweat. They had no idea what kind of corn their small plot of land would yield, and no clue what it would taste like as whiskey.

The time came to find out if all of their work was worth it—if it was, in fact, possible to make a good whiskey this way. Ann and Scott didn't yet have a mill of their own, so Glenn processed the corn and delivered it to Upper King Street. When it finally arrived at the distillery, they mashed it, fermented it, and hoped for the best.

Something weird happened. As it fermented, an oil cap formed, three inches thick. "We had never seen this before," Ann says. "We asked other distillers who had been in the business a lot longer and they said the most they'd ever seen was a small slick or a little puddle of oil."

But there it was, a layer of corn oil sitting on top of the tank. It had a distinct smell, one Ann compares to Banana Laffy Taffy. They blended it back into the mash instead of skimming if off the top, and sent it through the still.

A week later, they tasted it, unsure of what to expect. It was viscous, aromatic, and full of flavors they attribute to the terroir of Clemson's farm. Scott compared this to mushrooms, which are not all created equal because of the unique characteristics of where they are grown, like wheats, ryes, and corns.

"Place really does matter," he asserts, calling back to his time spent in Clemson's fields.

He wished they had grown fifty more acres. And soon they did exactly that, with the help of two additional farms in South Carolina's ACE Basin, which encompasses Lavington Farms, with which Glenn had put them in touch. Their goal was to get these three farms to grow enough Jimmy Red corn to sell.

In fact, the farms did grow enough of the corn for two small batches of whiskey. Ann and Scott laid it down for two years. They tried it, every so often, as it aged, siphoning a few ounces with a rubber hose from the barrel. Its flavor became more nuanced, more vegetal, fruitier, and nuttier.

They grew another three crops on the same farms while they waited for the first batch to mature, working just as much and hoping just as hard that history would repeat itself. When it was time, they put their bourbon into bottles and crossed their fingers that people would not only buy it, but would like it as much as they did.

The first batch was released for sale to the public in December 2016, almost three years after they started this project. Scott was at a farm picking up sugarcane juice for their agricole rum, leaving Ann to manage sales alone from the tasting room. They were selling the whiskey both at the distillery and online, and Ann began to worry that the silence would be deafening.

At 11:00 a.m., when sales officially began, the register started to beep. And beep. Between notifications that were appearing on her screen and the people who were coming through the door, Ann was inundated with orders. She could barely keep up.

Eleven minutes in, they had sold out. Ann had even sold the bottles that were meant for her and Scott. When he returned to the distillery, an unexpected sadness washed over him. He and Ann had organized their lives around this whiskey for so long, their trips to the farms steering the course of each day. Tasting it every few weeks was a highlight for them, and the anticipation of its release was consuming. And then, just like that, it was gone.

~

After Ann and Scott told me the story of Jimmy Red bourbon, we piled into their SUV and headed to the ACE Basin, which is where three river basins—Ashepoo, Combahee, and Edisto—converge, and is one of the largest undeveloped estuaries in the South. It's also where Lavington Farms was growing one of their Jimmy Red crops.

While other distillers have their own tales of risk and reward, few can drive just forty-five minutes to the farm where their grain is grown. Some can't even pick out the farms where their grains are sourced on a map. With food, and certainly with wine, producers tell elaborate stories about their ingredients. Curious consumers can visit goats grazing in a field in Northern California, grapes hanging on vines in Italy, or eighteen-year-old succulents growing on the hillsides of Mexico. But this doesn't happen with whiskey, mostly because distillers use corn that comes from industrial farms hundreds of miles away.

On the highway, as the road changed from asphalt to marshland, Scott rolled up the windows to save us from the thickening humidity. We pulled into a dirt road covered by a canopy of leafy green trees. A chorus of cicadas provided a soundtrack loud enough to hear through the sealed car windows. As the tree cover thinned, the road led to a wide field hidden by a neat row of shrubs.

"There's a rice paddy behind those bushes," Scott said, pointing to his left. "Jimmy Hagood, the man we're going to meet, is growing Charleston Gold rice."

Jimmy and Glenn had met at a conference a few years earlier, in 2010. Jimmy was talking about his family's farm, where he was the fourth generation, and how they didn't use pesticides or take extreme measures to control wildlife, preferring to let deer and wild hogs roam the land. They rotated their crops and understood the soil and paid attention to the weather. Historically, the land had been a rice plantation, as its location in the ACE Basin was perfect for a crop that needs plenty of flooding to grow. After listening to Jimmy speak about Lavington Farms, Glenn approached him to see if he'd be interested in helping to grow a plot of Charleston Gold rice. Glenn was well known in the Southern food community, and Jimmy was flattered. The next thing he knew, he was growing rice.

After passing the rice field, the road curved around to a clearing. There were square cabins, all painted white, each of their doors secured with padlocks, that lined each side of our path. This farm was a former plantation, and these cabins once housed enslaved people. Though Jimmy's great-grandfather bought the land in the 1930s, long after slavery had been abolished, his family kept the cabins in order to preserve the farm's history as a plantation. The road narrowed before leading us to a ranch house, its yellow paint faded from the sun. We parked under a tree, and as I exited the car I spotted a corn crop a few yards away.

A tall, slender, middle-aged man came around to greet us. Jimmy said hello to Ann and Scott, and as he introduced himself to me, I marveled at his accent, which was an interesting combination of a Charleston drawl and tidewater Virginia. He immediately escorted us to the corn, telling us about its progress since Ann and Scott's visit the week before. Jimmy Red wasn't historically grown here, so planting it on the farm was an experiment for everyone. Days earlier, an impending storm had forced him to harvest most of the crop, leaving us a single row to see on the stalk.

I watched Jimmy free a bright red cob from a desiccated stalk before I was invited to try it myself. I dove in, the brown husk crinkling between my fingers as I wiggled the corn loose. Up close, I was almost startled by its violent red color. The varietal is a relative of Bloody Butcher, which earned its name from aggressively cross-pollinating with weaker types of corn, turning them various shades of red.

"That's some fine looking corn, Jimmy," Ann said, beaming at the corn like a proud parent.

After we cleared the final row of corn, Hagood took us to a large open field where a truck with a deep flatbed sat alone, covered by a thick canvas tarp. He brought us over to it, placed his right foot on top of the heavy tire, and swung up to the top of the bed. He unsnapped the canvas, pulled it back, and jumped into the bed of the truck, disappearing from sight. He popped up a few seconds later, only his upper torso visible, with red kernels spilling from both of his hands.

"Scott, you've got to check this out!" he yelled down to us.

As soon as Jimmy exited the truck, Scott hopped in, slipping into the sea of kernels. "This is great, Jimmy!" he called from inside, wading among what would eventually be Jimmy Red Corn Bourbon. When Scott was done surveying their haul, we took turns climbing up the side of the truck, peering over the sides at the loose kernels.

While we stood in the grassy field of Jimmy's farm stretched out around us, a tractor rolled by. Lavington grows a number of crops, many of which are rotated to cultivate healthy soil. The ACE Basin is known for its biodiversity and rich nutrients. In the eighteenth and nineteenth centuries, a few large plantations were situated there, their main crop being rice. In the late 1800s, when the popularity of rice dwindled, the land was bought by rich sportsmen for hunting because wildlife thrived in the wetlands. It's now home to the Ernest F. Hollings National Wildlife Refuge, where the US Fish and Wildlife Service protects 12,000 acres that are home to endangered species such as the wood stork, as well as American alligators, foxes, feral hogs, and bobcats.

The other 23,000 acres of the ACE Basin are privately owned by people like Jimmy and his family. Being the fourth generation to own this land since the 1930s (which, it's important to remember, was historically worked by enslaved people), Jimmy knows how vital it is to protect the resources that have put food on countless tables. His farming practices—planting landrace grains that can't tolerate pesticides, rotating crops, and experimenting with small plots—reflect his interest in doing right by his land.

Standing in Jimmy's field next to Ann and Scott, pulling blood-red cobs from their sturdy stalks, gave me a glimpse of what sourcing cocktails could look like. I saw that whiskey does have a place in a healthy food system, that it is possible to make something full of flavor when you put the land first. Learning the story of the corn that ends up in the whiskey that I slowly sip at the end of a long day has made me appreciate this bourbon more, made it taste better each time I uncork the bottle.

Jimmy Red Straight Bourbon sent ripples through the industry, winning awards and garnering national media coverage. Indeed, it's part of a surging number of US craft distilleries—there were about 2,000 in 2020, up from roughly 200 in 2010. According to the American Craft Spirits Association (ACSA), sales of craft spirits rose by almost 30 percent just between 2017 and 2018. While craft distilleries remain a tiny part of the industry, and there's no guarantee that the "craft" designation means they produce sustainably, any shift to smaller, locally owned businesses bodes well for farmers. For ACSA, a craft spirit is defined as "a product produced by a distillery that values the importance of transparency in distilling, and remains forthcoming regarding the spirit's ingredients, distilling location, and aging and bottling process."

Another promising trend is farmers who are starting to do the distilling themselves. While moving from simply growing to distilling corn can involve a steep learning curve, this means that the people tending the fields are the same ones responsible for a spirit's taste—and that, in turn, means they are growing for flavor more than yield.

There's also a growing movement within the Black community to reclaim their historical contributions to whiskey production, with regard to both farming and distilling, and create space for modern enthusiasm around the category. There are Black writers like Leah Penniman, who have made important contributions to food and agricultural canon. Penniman's book *Farming While Black: Soul Fire Farm's Practical Guide to Liberation on the Land* aims to honor the legacy of her ancestors and to chronicle the fight for sovereignty over the land in the future. There have been archivists like Fawn Weaver, the historian who uncovered Nearest Green's story. She founded the Nearest Green Foundation, which will ensure that Green's legacy as a distiller will be forever included in the historical record. There are enthusiasts like Samara Rivers, who founded the Black Bourbon Society, which has 10,000 members and is growing. It is her organization's goal to bridge the gap between the spirits industry and African American bourbon enthusiasts across the globe. There are educators like Ashtin Berry, who co-founded Radical Xchange, a creative content agency that promotes and uses food and beverage as a conduit for conversations on equality, inclusion, and intersectionality. As part of this, she launched the Resistance Served symposium series, where bartenders, distillers, writers, and activists come together to discuss issues in the hospitality industry, including racial justice.

Interest in sustainability is growing every day in the food community. The James Beard Foundation, one of the most influential organizations in the world, has initiatives focused on the environment and regularly hosts panel discussions on a multitude of topics pertaining to sustainability. More and more people in distilling are finding their way to these events, opening the door for a community of others who are similarly interested in connecting food, beverages, and the earth.

Today, especially with the specter of COVID-19, it's hard to know if these industry trends will continue. If they do, just how sustainable, both environmentally and financially, will growth be? How much can craft distillers like Ann and Scott scale up their production while maintaining their values? Will they be able to find a way to distribute more widely, so sustainable spirits are available across the country or internationally?

To begin to answer these questions, I headed to a place where growth in the industry is putting tremendous pressure on both farmers and spirits producers: Guadalajara, Mexico.

CHAPTER 2

Agave

When I arrive in Guadalajara to meet with Pedro Jiménez Gurría, I know a few things. I know that Pedro is the founder of Mezonte, an NGO whose mission is to preserve traditional agave culture. I know that agave plants, part of the Asparagaceae family, are native to the Americas, particularly Mexico, and take years to mature. I know that for centuries Indigenous people have been creating all kinds of things from agave, including food, textiles, rope, paper, and mezcal. And I know that growing international demand for mezcal is putting tremendous pressure on the communities and environment that produce it.

Mezcal is an umbrella term, under which tequila is classified. This means that even though tequila is a type of mezcal, the two liquors have distinct differences. In 1949, the Mexican government issued a series of rules about how tequila was to be made, declaring that it could be produced only in the state of Jalisco and only with blue Weber agaves. Mezcal, on the other hand, isn't regulated as strictly; the government recognizes nine states that can make it, although it's produced in closer to twenty-six of Mexico's thirty-one states. It can be made from a number of different types of agave varietals, including those that grow in the wild.

While many people compare mezcal to a "smokey tequila," there is much more to it than that. Both spirits are produced by chopping and roasting the hearts of the agave, or *piñas*, in ovens—some underground—before crushing and distilling them. With mezcal, smoke is a part of the process, whereas with tequila, it's not. Distillers use various vehicles to smoke the piñas, such as agave leaves, mesquite, or wood, and often distill the mezcal two or three times to bring out the plant's flavors. Clay or bamboo stills are favored over more-modern copper ones. These traditional techniques have been passed down through generations of *mezcaleros*.

Agave—the plant, the spirit—is woven into the fabric of the country's landscape, economy, history, and social life. Moreover, agave culture symbolizes the relationship between the environment and the people of Mexico.

That relationship is being strained as people around the world have become enamored with agave spirits over the last two decades. According to Statista, close to 352 million liters of tequila were made in Mexico in 2019, the highest volume since 1995, which means that production increased by almost 237 percent over twenty-four years. In 2018, 5 million liters of mezcal were made, an increase of 38 percent over a five-year period. Those spikes have led to agave shortages, a problem exacerbated by the plant's long growing cycles. In turn, agave farmers have turned to industrial practices, often sacrificing water quality and soil health, and risking plant disease, in an effort to satisfy a thirsty global market.

The United States is the biggest export market for mezcal and tequila. I've heard a lot of opinions on these issues from white men, and a few women, who have a financial interest in selling agave in the United States; some have expert knowledge, though they are still tourists. And

I've heard the opinions of others who claim to know a lot about the industry but perpetuate problems that reek of colonialism. As a white woman who has visited Mexico a few times, I can't claim to fully understand the nuances of Mexico's agave culture. (This chapter isn't a comprehensive history of agave spirits and doesn't attempt to be one. For more, check out *Tequila: A Natural and Unnatural History* by Ana G. Valenzuela-Zapata; *Divided Spirits: Tequila, Mezcal, and the Politics of Production* by Sarah Bowen; or *How the Gringos Stole Tequila: The Modern Age of Mexico's Most Traditional Spirit* by Chantal Martineau.)

This is why I've come to Guadalajara. To better understand. And this begins with Pedro Jiménez Gurría. Pedro started Mezonte in 2009 because he loves mezcal, the people who produce it, and the places where they make it. The NGO works with small producers in a few states, mostly Jalisco, to bottle and sell their mezcal—at a fair wage. While he is guided by environmental and social consciousness, you won't hear Pedro on a soapbox, unlike several other mezcal companies with recognizable names, the ones whose ambassadors spout a white savior narrative at every talk they give. Instead, he's on the ground doing this work, regularly meeting with producers, running Mezonte's tasting room to educate people about agave, and occasionally traveling to the United States to talk about the industry.

The plan is to meet Pedro at a café near his bar, Pare de Sufrir, and then head down to southern Jalisco to the town of San Juan Espanatica to visit Arturo Campos, one of the mezcaleros who makes mezcal for Mezonte. At the café, there's room for just two small tables. Even though we're a few minutes early, Pedro is already there, chatting with the barista. We immediately spot each other; I've also got my husband, John, in tow for the trip.

Pedro is in his forties, his graying hair giving away his age. He is handsome and friendly, with a good sense of humor. He has a touch of artistry, which makes sense when you learn that he used to be in film production and takes many of Mezonte's photos himself. We sit and we sip coffee and we talk, John doing the heavy lifting as I ease out of my shyness. Though I talk to people for a living, it's usually about them. Very few of the people I interview ask me questions about myself. Pedro is different. He's conversational and interested in who we are. He asks about our time so far in Guadalajara, what inspired my research, and if we have any dietary restrictions. We catch up about the friends we have in common, the ones who originally put us in touch, and those who have taken trips with him in the past.

As we're talking, Zule Arias joins us. She is the manager of Mezonte's tasting room and has worked there for four years. She's somewhere between her late twenties and early thirties, slender with black hair, a round face, and an artsy vibe. While she waits for her coffee, she pulls a camera out of her bag and sits at the counter behind us, tinkering with its settings.

We've actually met her before, about a year earlier when John and I were in town and our friends took us to the tasting room. She might recall our friends, who work at restaurants and bars in Guadalajara, but there's no way she remembers us. A lot of people come through Mezonte to taste mezcal. This is how she initially regards us—just people passing through. She loves mezcal just as deeply as Pedro does, and it's clear that she wants to protect it, along with the people who make it. Perhaps that's why she seems skeptical of us, foreigners, in her city, here to talk about her industry. But she softens over the course of the day, more with each question we ask and each joke we make. By the time our trip has ended, she is the one wisecracking.

Once we're all sufficiently caffeinated, we follow Pedro to his car. There are a few plastic carboys in the truck behind us for the frequent treks that Pedro makes to visit the mezcaleros. The drive to San Juan Espanatica will be a little over an hour, so I take the opportunity to ask Pedro and Zule some questions about their work. Pedro, who loves food, takes the opportunity to eat, making time along the way for what he tells us is a gastronomic tour. But before we stop for our first round of tacos, we talk.

"How did you originally get interested in all of this?" I ask from the back seat.

"By drinking mezcales," he replies with a smile. He says that he has been drinking mezcal since he was eighteen, back when he couldn't be picky and most of what he drank was cheap. "But every time I drank mezcal, it was a different experience. I started to ask myself, 'What's with this mezcal thing?'"

He was living in Mexico City then and didn't have much access to mezcal. He started working in film production and was doing documentary and commercial work, which took him all over Mexico. Whenever he traveled, especially to Oaxaca, he would bring bottles back. He would ask locals what type of mezcal they had there and occasionally got to meet the producers. Eventually, he moved to Guadalajara after he met his wife, Mónica Leyva, in 1998 (though they didn't get together until 2004).

"When I moved here, there weren't any good mezcales available. I was in pain. I was suffering," he says, giving us a good laugh.

Pedro continued traveling for work and amassed a solid mezcal collection. He would often host tastings for his friends, who were more familiar with tequila, which can be produced only in Jalisco, making it readily available in Guadalajara. He released a documentary, *Viva*

Mezcal, in 2012, which was about contemporary mezcal culture; today, people are more aware of the importance of agave's place in shaping Mexico's national identity than ever before. His love for agave was palpable, and his friends jokingly told him that he should open a *mezcaleria*, planting the seed for Pare de Sufrir.

"I would think, 'Oh, maybe that would be a good way to have access to good mezcales for myself.' Very selfish, but maybe that's the way," he recalls.

Pedro opened his mezcaleria, Pare de Sufrir, which translates to "stop suffering," in 2009. He stocked the bar with bottles from producers whom he already knew. It snowballed from there as recommendations rolled in for other producers. His network grew, and he's been working with many of the same mezcaleros for more than ten years.

In the beginning, the bar was open only for tastings a couple of days a week. As people became more interested, he added music. Now, it's open most days, filling up with people on busy nights, especially when there's a popular DJ spinning records.

I'd heard a lot about Pare de Sufrir before I first visited. Its reputation for a vast mezcal selection and good music travels across borders, making it an international destination. It's a square, open room with murals painted on the wall. A DJ booth sits in front of one, an orange-striped city bus named "Dina" with blue skies above. Another mural of giant agaves crowns the wooden bar shelf lined with mezcal. The menu spans the top half of one of the walls; it's written in colored chalk, with photos of a few of Mezonte's prominent mezcaleros pinned underneath. A disco ball hangs overhead, fractured light dancing on the walls. There are no frills here, only good liquor, bright colors, and loud music. They don't serve cocktails either, just neat pours and beers. Unlike the highly stylized bars I'm used to in San Francisco and New York, Pare de Sufrir is unapologetically itself. Its popularity led to Mezonte.

"For me, it was essential to have a place where you could have more information available to appreciate what agave culture is," Pedro says. "We decided to open another space so there would be a link to what you're drinking at Pare de Sufrir. If you want to know more and have a conversation, Mezonte is the place." Mezonte's hours don't overlap with those of the bar, so there's no confusion about the difference between the two places.

Mezonte's tasting room is under the direction of Pedro, but the experience is created by Zule. She has a deep knowledge of mezcal, down to minor details of each producer's story and techniques. She answers questions earnestly, regardless of the topic. "What is the difference between tequila and mezcal?" gets the same weight as "What type of agave grows in the region where this mezcalero makes his mezcal?"

A while back, during my first visit to the tasting room, Zule poured us *copitas* of different mezcales, reaching for bottles of various sizes behind her. She poured us *racilla* and *bacanora* from a few regions. The pair that stood out the most were made by Don Lorenzo Virgen and his son, Tomás, who are from Jalisco. They distilled agaves from the same field with the same equipment, using the same methods, but they tasted completely different. The hands that create each mezcal make it unique, with its own distinct flavors.

Of course, it's not just the hands that make mezcal. It's also where the agave grows, each region reflecting various characteristics of flavor, of technique, of the environment. The landscape—the composition of the soil, the elevation, the sun, the rain, the wind, the local pests, the surrounding crops—imbue a particular taste, just as terroir does for wine. These biological qualities combine with the techniques distillers use to highlight agave's flavor. And this is where human artistry creates the magic of distilling. It's why a son can make a mezcal from the same agave crop as his father and it will have a different profile.

Zule's tour through the Mezonte was one reason I wanted to see how these mezcals were made. While all the bottles share the Mezonte label, the names of each producer and region are clearly attributed, along with the production method, the type of agave, and the number of liters in each batch.

During our drive, I ask Pedro about his relationships with the producers. "What's the process of getting them to work with you?"

"It was the same thing as doing documentaries. You have to start knowing them before you start working with them. We had conversations to see if we're on the same track," he explains.

Pedro talks extensively with all the producers about how much mezcal they are currently making and their potential to generate more. If sowing more agave plants will put too much pressure on the land or make the crops vulnerable to disease, Pedro won't push production volume. Sometimes, this means that producers will make less than they want, but the trade-off is that the land will be able to support agave crops for years to come.

"We want to represent Mezonte. It's more about the people who care about tradition and understand the cycles of the lands, even the moon cycles, which are constantly used in farming in rural communities across Mexico. Producers like Don Lorenzo harvest their agaves in the full moon. They are doing a great job with the ecosystem. They are keeping it small and it's really clean and really round, and that's what we're aiming for," he says.

In many cases, Pedro is paying producers more than what they are used to making, which eases their desire to distill more mezcal and push the limits of the environment. In addition to sustainability, his biggest priority is the producers' financial livelihood. His third criterion is geography. He prefers to support mezcaleros from areas such as

Jalisco and Durango and Michoacán that are often completely ignored. "Besides, Oaxaca doesn't need more promotion," he says, chuckling.

"When you're talking to producers about environmental issues, are you hearing anything come up that concerns them the most?" I ask.

"The shortage of agave is the main one," he says.

There are over 200 agave varietals, 150 of which grow in Mexico. Agave has been a Mexican staple for thousands of years. But in 1949, when the Mexican government mandated that tequila be made from only 100 percent blue Weber agave, it set off a cycle of boom and bust.

The tequila industry grew in the 1950s and '60s, which, on the heels of these new quality standards, quickly created an agave shortage. Demand for tequila went up as availability of blue Weber agaves in Jalisco plummeted when the plants were harvested.

Tequila producers pushed back against these standards, knowing that their distilleries wouldn't be able to survive the shortage. The Mexican government eventually relaxed the regulations in the 1970s, allowing tequila to be made with a smaller percentage of blue Weber agave.

Enter *mixto* tequilas. Since tequila no longer had to be made solely from blue Weber, things like corn and wheat were used to supplement the recipe. These mixtos weren't as palatable as tequila made from 100 percent blue Weber, so companies added flavorings and sweeteners to mask the taste.

As mixtos were entering the market to alleviate the pressure of the blue Weber shortage, food production everywhere was being industrialized, and tequila was not immune. Big businesses, seeing the rising demand for tequila and their potential to profit, bought out small agave distilleries all over Jalisco. They introduced new technology, like column stills that could yield a higher volume, and diffusers, which shred agave plants and use steam and chemicals, like sulfuric acid, to extract sugar—up to 20 percent more than by traditional methods. Column stills and diffusers, especially when used in concert, strip out many of the flavors that give agave spirits their character, leaving it with little terroir.

During that 1960 shortage, farmers also sought to benefit from the high prices created by scarcity, so they planted more blue Weber agaves. When these plants matured six to eight years later, an abundance of blue Weber agave was suddenly available for sale, driving down the price. This was exacerbated by tequila *coyotes*, the intermediaries between independent agave growers and large distilleries. These coyotes had the power to drive down the price of agave that is paid to the smaller farmers who can't sell directly to the big companies. Therein lies the rub: during a surplus, farmers can't get a good price for agave so they don't want to plant more, though that would help them plan for the next shortage. In some cases, farmers have abandoned entire fields, unable to sell what they've grown, unable to reap any reward from a nearly decade-long investment.

This economic roller coaster gets even bumpier when farmers turn to monocultures to ramp up production of one specific crop. Things that traditionally grow quite well in parts of Jalisco, like maize, beans, and squash, are displaced to make room for agave. There's no more intercropping, so soil health deteriorates, and without other crops to act as a buffer, the plants become more susceptible to disease and insect infestation. All the while, there's a decrease of the beneficial species like birds and beetles. And since most blue Webers are grown from the same clone, an entire field can fall victim to viruses and bacteria because they are genetically weak, allowing disease to spread like wildfire. This often causes farmers to ramp up their use of agrochemicals that target one or two types of pests while killing dozens of other good ones in the process. Some of the chemicals that farmers use as pesticides are extremely toxic to the pests and people alike. The herbicides that are sprayed on the land end up stripping the soil of the nutrients it needs to be fertile, thus depleting the vegetative cover, which leads to erosion.

The cycle continues, and it isn't exclusive to blue Weber. In the 1980s, there was a wild agave shortage. Agave prices shot up, and many small producers couldn't afford to buy wild agave. In Oaxaca, the region where most of Mexico's mezcal is made, over 1,000 *palenques* (distilleries) were forced out of business.

Then, in 1994, the Mexican government established a Denomination of Origin (DO) for mezcal, much like it did for tequila back in 1949. Originally, five states were designated; today there are nine: Durango, Guerrero, Oaxaca, San Luis Potosí, Zacatecas, Guanajuato, Tamaulipas, Michoacán, and Puebla. Jalisco is only under the DO of tequila, since 1974, and racilla got a DO in 2019. Because of these official designations, wild agave is now subject to similar pressures as blue Weber and the mezcal industry is staring at the same boom-and-bust fate as tequila.

After the DO was established for mezcal, the threat of another agave shortage loomed large. Like the tequila producers before them, mezcaleros looked to industrial models, using column stills and diffusers, monocropping, and spreading pesticides and fertilizers to boost production.

And as with blue Weber, these measures backfired. An unusually cold winter in 1997, followed by a summer infestation of weevils, devastated the agaves. There were no other crops in the fields—no corn, no wheat, no strawberries—to distract the pests or buffer the spread of bacteria. By 1998, 25 percent of Jalisco's plants were in trouble. Entire fields were wiped out in a matter of months.

The blow was swift. Fifty percent of agave crops were lost. The price of agave soared, hitting an all-time high in 2002. Thirty percent of

Mexico's distilleries shuttered. The country's production dropped by 20 percent, while tequila prices climbed by 20 percent. All this as America's cocktail revival was taking off, calling for more and more agave spirits.

The industry was in crisis. People demanded that farmers heed the wake-up call and alter their practices by planting different agave varietals and crops in their fields. Others wanted the government to step in and create safeguards, like a ceiling and floor for the price of agave.

The government did act—but unfortunately in ways that only ramped up the pressure on small growers and distillers. In 2005, the Mezcal Regulatory Council (CRM), an agency that regulates the production of agave spirits, started certifying mezcal. The application process requires a company to send in samples so that their ethanol and methanol levels can be laboratory tested to ensure that they fall into the required range. Certification doesn't have anything to do with how agaves are grown or how mezcal is made. It's an expensive process, similar to agricultural certifications in the United States, so only about 10–20 percent of mezcal producers pay for certification. The companies that can afford it all mass-produce mezcal, which privileges damaging industrial practices, supports the boom-and-bust agricultural cycle, and strips agave of its terroir.

One of the ways the CRM tries to regulate mezcal is by issuing a NOM, or Norma Oficial Mexicana, which "establishes the technical specifications and legal requirements for the protection of the Appellation of Origin." The government came out with a Denomination of Origin (DO) for mezcal in 1994 to officially recognize the state in which it was made, thereby permitting only a fraction of places that have been making it for generations to call their spirits *mezcal*. The states that fell outside of the DO aren't officially allowed to make agave spirits that are called *mezcal* and must be called something different.

Then, in 2011, the tequila regulatory council, or the CRT, the organization that regulates tequila's appellations, drafted NOM-186. It was written by representatives from the appellation of origin, which indicates where something was made, for tequila, mezcal, and bacanora. The legislation was proposed under the guise of adding clarification on tequila and mezcal labels so that the industry could be more closely regulated. But it would drive up sales for big companies who exported to countries like the United States, the largest market for agave spirits outside of Mexico.

NOM-186 attempted to change the name of all agave spirits from whatever they had always been called—*mezcal, bacanora, racilla*—to "spirits distilled from agavacea." Around the same time, the CRT tried to trademark the word *agave* for private use.

"It was basically their first attempt to 'own' the word *agave*," Pedro says, adding that this was absurd and illegal.

There was also more to the proposal. It defined what could and couldn't go into a bottle of mezcal, much like what was done with tequila back in 1949, by limiting the type of agave that could be used to make it. Later, in 2015, NOM-186 was expanded when NOM-199 was drafted by the CRT and the CRM as an attempt to prohibit the use of the word *agave* for all agave spirits that weren't under a DO. NOM-199 would only recognize mezcal made in eight states (not the nine it currently recognizes). This meant that the mezcaleros whose families have been making mezcal for generations outside of those eight states (of the twenty-six states that produce mezcal) could not export their spirits commercially. Producers wouldn't be able to use the word *agave* even to describe the raw material that they used for their distillate. If they did want to sell their mezcal, they'd have to label it as *aguardiente*, which colloquially translates as "fire water." They also wouldn't be able to label their spirits as "100 percent agave" because they'd be using wild agave varietals that weren't recognized by the two NOMs.

If NOM-186 had passed, mezcal would have faced many of the same environmental pressures as tequila. Growing only specific types of agave in a handful of states would further encourage harmful farming practices, like mono-cropping. There would be more boom-and-bust agave surpluses and shortages, weakening the plants and the land, all for the sake of a few big businesses. Small producers would be forced to close their palenques, and generational knowledge about craft, technique, and land management would be lost, making it impossible to recover in any meaningful way.

"By doing the same thing that they have been doing for generations," Pedro says, "suddenly they are illegal. That's why we are always so mad about these legislations. We've been through this for at least the last eight years. With every proposal we tell them, 'You should consider this, and this, and this. You should consult and hear what the producers need.' But that hasn't happened."

In the wake of this proposal, people all over Mexico sprang to action, recruiting allies from the United States. Pedro was one of the people involved in fighting NOM-186, and he spent much of 2012 educating producers on its dangers and explaining why the stakes were so high. Advocates, small business owners like Pedro, academics, researchers, scientists, writers, and bartenders came together to spread the word and circulate a petition, which eventually made its way to the United States. I attended a talk about the issue in San Francisco and signed the petition. So did legions of other bartenders and bar owners, who threatened to boycott agave spirits if the NOM passed. As a result of all of this collective work, NOM-186 was successfully defeated.

"The whole community of bartenders across the world signed a petition to say no, and that stopped the government from going through with it," Pedro says. "People in the US represent money, so when they say no—especially with the word *boycott*—then suddenly the tequila industry listens. It's just a few big tequila companies, but they have a lot of control and power. Those companies use it as a marketing tool and said that they stopped it. We were like, 'What? What are you talking about? You're the one who started it.'"

Though NOM-186 was defeated, the threat of a similar regulation is always in the back of Pedro's mind. The big spirits companies in Mexico have a lot of power.

"These guys producing these legislations are the same guys who are monopolizing the whole market," Pedro says. "So it's a thing that we won't stop fighting. At the same time, we have to build some other ways, some outside ways, to protect producers and protect the culture around the spirit. It's so tiring."

When I comment that the process has to be exhausting, Pedro replies, "It is. But the good thing is, we've built bigger networks with people who are concerned. You start to know a lot of people who are already working on these things. People who are conscious and aware and committed to these spirits or agave production or the ecosystem. Suddenly we're all at the same conference or writing letters to the government together. We get to spread more information, and people at all levels, from the producers to the consumers, are more aware of what is happening. So hopefully that will grow and there will be a point where we can stand together and say, 'No.'"

⌒

Pedro met Arturo Campos and his father in 2014 after a friend of his, an anthropologist, tipped him off about their mezcal. They were making mezcal from a type of maguey called *cenizo* as opposed to the traditional *espadín*, which has become the most common varietal because of its high sugar content and relatively short growing cycle. It tasted good. Over the last ten years, as the popularity of agave spirits has grown, Pedro has seen many producers whom he respects start to make mezcal without much character, catering to what they think their audience wants. Something has been lost along the way.

"And then I tried Arturo's mezcal. Once again, I found something with character. It tasted delicious," Pedro says. Arturo and his father started working with Pedro immediately.

With that, we're here. "Welcome to San Juan Espanatica, Land of the Carnitas," Pedro exclaims. "It's basically two whole streets, this whole town. There are probably eighty different businesses, and sixty of them sell carnitas. Also, if you want some corn, we can have that as a dessert."

We head for Pedro's favorite carnitas stall, the one that was recommended by Arturo. We're standing around a giant cast-iron cauldron, picking out our protein, when a tall, round man in a gray polo shirt waves at us as he's crossing the street.

"Hola!" Arturo greets us, flashing us a big smile.

We eat and we talk and, when we're finally full, we head to Arturo's *tachica* (which is called a *palenque* in other regions). It's a few minutes away from the center of town, off the street and hidden from view of the road. Arturo shows us around. We're a few weeks early for harvest, so his tachica has been quiet for a few months. I've seen many active distilleries, so I'm more interested in the equipment that Arturo is using to make his mezcal, the kind that brings out the flavors that give it a unique terroir.

It's a small operation, with one area where he crushes the agave, one pit where he roasts everything, one stone tank where he ferments each batch, and one clay pot where he distills it into mezcal. Though he could handle seven tons of agave, he usually distills only five tons because that's what fits in his stone fermentation tank. If he distills any more than five tons, he has to ferment it in plastic, which affects the flavor.

Pedro leads us to the area where Arturo crushes the agave. "*¿Todo bien?*" he asks Arturo before he pulls back the blue tarp, where spiders have set up residence during this dormant period.

"*Sí. Sí, sí,*" Arturo says, and then tells us about his distillation process. He speaks fast to Pedro, but slows down a bit when he looks at me. Though I get the gist, my Spanish is rusty, and Pedro is kind enough to translate so that John and I can understand everything properly.

The process goes like this: the agaves grow; they are harvested each fall when they're mature; they get crushed, roasted, fermented, and distilled. Fermentation usually takes between ten and twenty days. While most mezcal producers crush the agave, put it in the fermentation tank, and wait anywhere from one to six days before adding the water, Arturo adds water as he puts the agave in the fermentation tank. First, he fills it with agave to about five inches below the rim. He then boils the water and blends it with the agave, tasting the mixture along the way and adjusting his ratio based on its flavor.

"If it's really sweet or without taste, as they call it, he keeps blending it and adds more water or more agave, depending on the flavor," Pedro says. "The whole fermentation process takes about twelve days, depending on the season and the temperature." The weather makes a big difference, as everything is done outside. There's no fancy building with regulated climate control here.

Arturo, who often distills with his father, typically uses two tanks. He tries to balance the timing of each fermentation so that neither gets overdone. Arturo prefers to distill everything at once and includes the bagasse, or the fibrous by-product of the fermentation process, which is something most distillers don't do. After these two batches are fermented, Arturo distills them, twice, using a clay pot.

"They add the firewood on the bottom," Pedro explains. "It starts to evaporate in the clay pot, which heats the cooler, and then it starts to condense and drips into these spoons. They have channels that they travel through and hit this agave leaf, which is like a funnel."

Both the clay pot and spoon have a big impact on flavor. So does distilling outside, where there is a lot of natural yeast in the air, a water pump nearby that allows them to add cool water throughout the process, and wood from nearby to power the furnace. This is a traditional way of making mezcal, and it's part of what Pedro wants to preserve through Mezonte. But it's not without its challenges, like modernization of equipment.

"Since Arturo is working with clay, the producers of clay aren't working it the same way they used to," Pedro says as Arturo nods in agreement. "Now they make these clay pots in two stages, and that creates a crack in the middle because it's not sealed at the same time. Because mezcal is connected to the artisan who makes clay pots, that's decaying, too. There's a lack of people who are still doing clay pots well."

Arturo has been distilling for well over ten years, but he has been using his current tachica for just the last decade. He learned how to make mezcal from his father, who learned from his own father. Arturo is his family's fourth-generation Jalisco distiller.

It was also from his father that he learned how to grow and harvest agave. Right now, Arturo grows agave in three different areas. They rent the land, but they are trying to buy it. One of these fields is our next stop.

As we pull away from the tachica, Pedro tells us that distilling has been a persecuted activity ever since the time of the conquistadors. When Mexico was a Spanish colony, the Spanish government would tax mezcaleros and sometimes destroy their fermentation vats. Though this stopped when Mexico became independent in 1810, the Mexican government stepped in with its own taxes and intimidated mezcaleros into giving officials a cut of their profits. Though this practice ended a long time ago—at least one generation earlier—other actors are using the same tactics to control agave commerce. The Consejo Regulador del Mezcal, a government entity, is giving mezcaleros a hard time for making mezcal, pressuring them to turn greater profits instead of protecting the quality of agave, traditional methods of production, and cultural and biological richness.

"This is called *barranca*," Pedro says, explaining that mezcal goes by this name in Jalisco so it won't be taxed the same way as tequila or mezcal. Each region has developed a different name for mezcal to shield it from government or illicit persecution. "Producers will tell you, 'No, this is not a mezcal. This is not a tequila. This is not a racilla. This is barranca.' I love that. It's a cultural thing, depending on each region."

Along the short drive to Arturo's field, we stare out the window, taking in the fields that surround us.

"Those are all avocados," Zule says, pointing to rows and rows of trees. "This area will get flooded, and there will be a lot of wildfires near the volcanoes. They wipe out the fields so they can plant later."

"Like slash and burn?" I ask.

"Yes," Zule confirms.

We pass fields with young blue Weber agaves, which grow along the road. We pass strawberry fields. We pass corn crops.

The road winds around a small mountain and quickly turns from asphalt to dirt. We pull over onto the narrow shoulder. Across the street is a field full of agaves, their spiny leaves poking through the fence. I pause to swap my sandals for socks and tennis shoes, the earth ahead being a little too uneven for flimsy footwear. Pedro and Zule wait for me as I change into my shoes while John disappears between the plants behind Arturo.

As I'm bending down to tie my shoelaces, I notice that there are ants everywhere. They are big with fat, round bodies the size of my pinky nail, and a deep red color. They are marching toward mounds of porous ant hills that dot the dirt.

Pedro tells me the ants are actually a sign of healthy soil, aerating it and making it fertile for the agaves. Based on all of this ant activity, the soil looks quite robust.

I sling my backpack over my shoulder and nod to Pedro, who leads us up the sloping hill. Zule falls behind us, taking her time to snap photos. Pedro points out a few plants that have already had their *quiotes* (flowering stalks) cut, which many feel makes the flavor of the agave richer. The leaves of the plants are tough, their tips sharp. This is not the meticulously manicured blue Weber field where tourists pay to shuffle around in Jalisco. Arturo maintains these plants himself. A few plants have had their tips strategically cut, creating an Arturo-size path to slide between leaves.

"How old are these agaves?" I ask

"These are probably around seven years old," Pedro says as he assesses the plant. I follow him farther up the hill, maneuvering between the leaves. "Just be careful." I can hear John and Arturo laughing in the distance. The grass crunches beneath our feet.

There are several different varietals of agaves in Arturo's field. Pedro points a few out. "This one is used less because it doesn't have the strength," he says. "Almost all of them are cenizo. And the really big one on the end of this row probably *verde*."

He calls to Arturo. "This is verde, *¿verdad?*"

"*Sí!*" Arturo calls back. It's been there for five years, but it was replanted when it was two, making it a seven-year-old plant.

"This is one of the things to look for in different types of agave—the thorn. This cenizo is from the verde. And the tip, that's another way of identifying them," Pedro says.

"Because they vary in length and width?" I ask.

"Length. Width. Strength. Separation. Direction," he explains.

"Was the quiote cut?"

"Yeah, when the quiote is about the size of the leaves, it's the right moment to cut," Pedro says. "After that they can leave it for three to six months. Arturo might leave it up to five years. Those are plants that take thirteen years to grow and still take another five to mature. Once again, people are looking for short-term fixes."

He explains that when the quiote is cut at thirteen years, mezcaleros might wait anywhere from six months up to four or five years to harvest the agaves. Though the plant is already mature, this is called *capones*, which translates to "castrated"; cutting the quiote forces the sugars inside the agave to become more concentrated.

"Do you have to irrigate them?" I ask. "Or do you rely on natural rainfall?"

"Natural rainfall," Pedro says.

"That's what keeps them fighting," John adds.

"Yes, exactly," agrees Pedro.

Since agaves are dry-farmed, droughts have a huge effect on their growth. Groundwater is also important because the root system absorbs water from the earth. As farming in Mexico has increased, thirsty crops destined for export, such as avocados, have started limiting the water available to agave.

Once the agaves have matured, Arturo harvests them himself. Sometimes he hires a couple of people to help him cut the leaves, unearth the agave, trim the piña, and haul the plant down the hill to his tachica.

"The whole process, from harvesting to distilling, will take between eighteen and twenty-two days. If he has two other people harvesting with him, it will take three days to harvest five tons of agave. And then they put it in the pit for four to five days and crush it one more day, and then the fermentation and distillation," Pedro translated for Arturo.

"That's hard work," I remark.

"Yes," Zule agrees. "Even for the harvesting, it is hard. You have to be really strong."

Arturo looks down the hill at his field. He tells us that it brings him a lot of joy to see what he's grown, given the effort that it's taken him to get here. Coming from having nothing, he's created this. It might not feel like much, but every step counts.

"I can testify to that," Pedro says. "They didn't have the land, they didn't have the agaves. They've done a lot of hard work. It makes them really proud."

Pedro says that he thinks that most producers love coming to their agave fields. "Who wouldn't?" says Pedro. "I feel proud standing here, and they aren't even mine."

"They are beautiful," I tell Arturo.

"*Están bonito*," Pedro translates.

"*Sí*," Arturo says, drawing the word out as he surveys the plants. "*Sí.*"

We make our way back down the hill, weaving through the plants. We pass the big agave on our way down. John asks me to take a photo of him and Arturo in front of it. They strike a few different poses, including one with their arms stretched over their heads, the agave leaves taller still than either of them. They smile and laugh and hug. We're all quiet as we make our way back down, taking in the field.

For as much as I've read or heard about mezcal, there's nothing like actually visiting an agave field, especially when it's cared for by the person who is also harvesting and distilling the plants himself. The hard work that goes into making mezcal feels romantic when you're walking through the fields, watching the mezcalero beam at what he's created with his own two hands. But it's so much more than a luscious, pastoral scene. It's history, culture, heritage, economics. It's the environment. Agave is the linchpin, and the stakes are high.

CHAPTER 3

Gin and Vodka

I am a person who makes New Year's Resolutions. I don't remember most of them, but I do remember one I made in 2010: Drink more gin. I was in graduate school when cocktails became a bigger part of my life, owing mostly to my bartending job in Brooklyn. I dove into cocktail history, reading a handful of books—mostly to keep my job interesting, but also to get better at it. Recipe after recipe, classic and contemporary, called for gin.

I was not much of a gin drinker before that. I'd never gone through that phase in college because I liked whiskey better. Sure, I drank some gin over the years, but the piney taste put me off. Whenever I'd try it, all I could think of was the Pine-Sol cleaner that my mother used to polish furniture.

The fateful year I decided I needed to be a more adventurous drinker, gin was in my crosshairs. The dawn of that new decade brought what seemed like hundreds of distilleries to the Northeast, especially Brooklyn. Gin was unavoidable, and once I started paying attention, my options were limitless. It took me a few tries to find my cocktail of choice. But soon, Negronis and Bijous crept their way into my frequently ordered drink rotation. I haven't looked back.

It wasn't until I moved to California in 2013, with my feet tucked underneath a barstool, that I started noticing one brand in particular: Leopold Bros. One day, my interest piqued after seeing their gin on shelf after shelf, I ordered it in a Negroni. The drink was nuanced, balanced, and layered. There was no taste of Pine-Sol. There were no memories of dust.

As time wore on, I found that the Leopold brothers were creeping into more of my conversations. Friends who visited Denver went to the distillery, returning with bottles of their gin as gifts. Distillers who knew them talked about their work during our interviews, citing them as industry leaders. Bartenders I respected used their products. Magazines I read kept featuring them in articles.

One of these articles profiled a lauded San Francisco bar owner, Thad Vogler. At the time, Thad owned several bars, including the James Beard Foundation Award–winning Bar Agricole. I had come to know Thad while doing an oral history interview with him. Over the course of the hours that we spent talking, he went into detail about how spirits are agricultural products, and why we should treat them that way. Bar Agricole, more than any other bar in the country, embodies this philosophy. They refer to themselves as a "farm bar." They source all of their products meticulously. They want their producers to be environmentally conscious. And they want their drinks to taste good. While Bar Agricole carries mostly European-produce spirits, Leopold Bros. is among the few spirits they stock that are made in the United States.

After talking with Thad and drinking the Leopold Bros. gin, I got curious about them. I learned that sustainable distilling practices had been central to their business model from the very beginning. Their systems are designed to reduce waste—water, energy, transportation, and food—and they support local farmers and flavors. They, too, treat drinking as an agricultural act.

And with that, I booked a flight to Colorado.

~

I arrived in Denver on a clear, bright day. Driving from the airport toward the city limits, I passed small solar farms, their black panels tilted toward the sky. Once I made it downtown and began walking the streets, I regretted not wearing sunscreen. The buildings and construction equipment shaded me from the sun, but every few blocks it found a crack in the shield. Passing through those heat patches reminded me of swimming in the ocean and feeling a warm current pass. It wasn't until those moments, the light radiating my face, that I began to understand the power of solar energy here, and why the Leopold brothers use it to their advantage.

The Leopolds are brothers Scott and Todd. Scott, the older of the two by a couple of years, went to college at Northwestern University, where he earned dual degrees in economics and industrial engineering. He later got his master's at Stanford University in environmental engineering. Todd wanted to learn how to make beer, and he pursued a degree in malting and brewing from Siebel Institute of Technology in Chicago. After graduating in 1996, he shipped off to Europe to train at the Doemens Academy in Munich, where he focused on lager. While in Europe, he took the opportunity to apprentice at several breweries and distilleries.

The two men got their start as a company in 1999, after years of planning. They set up shop in a warehouse in Ann Arbor, near the University of Michigan campus, with a brewery on one side and a taproom on the other. They were making a handful of beers: ales, ambers, and stouts. They used organic ingredients, and Scott implemented a system to manage wastewater and brewing by-products.

The brothers had grown up in an eco-conscious household. Their father, Bob, was influenced by Aldo Leopold, the famed conservationist who wrote *A Sand County Almanac* and shared the family name. Bob wanted to follow in Aldo's footsteps. He worked as a landscape architect for the federal government, folding the surrounding flora and fauna into the design of national parks and Bureau of Land Management land. He took what was there, highlighted it, and gave each park a unique identity.

This made an impression on Scott. As a teenager, he was full of youthful optimism, determined to change the world. Like his father, he was interested in the environment. He wanted to hand down something better, healthier, to the generation of people who would come after him. When it was time for him to apply to colleges, he picked engineering because he thought the biggest threat to the planet was industrial waste, and he wanted to fix it. In Chicago and Palo Alto, he learned how to design systems that would eliminate or reduce waste from production facilities.

For six years, Scott worked as an environmental engineer. "My specialty was in pollution prevention," he says. He worked with a myriad of companies that made everything from chairs and crayons to electronics. Many of the companies who hired him were on the Fortune 500 list, and many were in violation of environmental regulations. So he took their systems and fixed them. It was a win–win for them, and the environment.

Scott was in the middle of his career during the mid-1990s, a period when breweries were opening in cities, big and small, across the country. Brewing beer is water-intensive; water is used to cook the ingredients, cool the equipment, and to clean, which is the most wasteful part of the process. It takes anywhere from three to eight gallons of water to make a single gallon of beer.

"There were all of these breweries popping up, particularly in small towns," Scott explains, "and they were one of the main contributors to the wastewater stream." These breweries were generating a lot of waste, and their county and municipal wastewater systems were routinely overloaded. He started getting calls from breweries and city agencies, asking for his help. "Back then, for every glass of beer they produced, they would generally create anywhere from fifteen to thirty glasses of wastewater. So the phone calls would come, and I started working with breweries."

Scott designed a system to help breweries cut down on their water consumption and treat the wastewater by-products. The efficiency was good for both the environment and the breweries' wallets, since they were spending less on utilities. It was through this work that Scott was able to design a system that he and his brother would later use at their own brewery and distillery.

At the same time, Todd was in school learning how to brew beer. One weekend, when both brothers were visiting home, Scott from Kansas City and Todd from Chicago, their conversation drifted to ideas for a business they could start together, like a brewery. And then it strayed into uncharted territory: a sustainable brewery. Scott was tired of advocating for other people to change their habits. He wanted to build a model for himself that other people could follow. Todd, eager to set his own rules for making beer, felt the same way.

They spent what Scott calls a "couple of years in designing and planning and financing." What this means is they had to learn how to channel their skills collaboratively. Todd told Scott what he needed to brew, and Scott drew up the plans.

"My brother knew how to brew," Scott says, "and I knew how to engineer. You need both to be effective." There were no textbooks to consult, no models to follow. "If there was anyone else doing something similar," Scott chuckles, "I didn't happen to know them." They had to start from scratch, with nothing other than their own experience and philosophy to guide them.

Financing their dream was another story. Banks haven't always been kind to companies that make alcohol, a throwback to Prohibition-era notions about people who drink. What's more, they were both young—and looked it—and had never started a business. Loan officers were less interested in their experience and more in their resources, questioning if there was a market for this kind of thing.

After being rejected by bank after bank, they had to reformulate their business plan. Part of their revised approach was to bring in their father, hoping that Bob's gray-haired wisdom would make them seem more seasoned, more serious. After more rejections, they decided to give it one more shot. This time, it worked: the bank gave them a portion of the money they needed to start the project. Construction started in 1998, transforming the Ann Arbor warehouse into the brewpub with tables and space for live music. They served their first pint in October 1999, just in time for the fall.

Just a couple of years later, in 2001, the brothers grew the business into a distillery. Todd was able to put what he'd learned during school and his apprenticeship into practice, and Scott got to apply the same waste-reducing systems that they were using for the brewery to their distillery operation. They ran things like this until the mid-aughts, when they decided it was time to move back home to Colorado. They closed the brewpub so they could focus on the distillery, moving their equipment to a small temporary workspace in an industrial part of northeast Denver.

Today, the Leopold Bros. distillery is outside of Denver's city limits, a fifteen-minute drive from the airport. I passed coffee shops, strip malls, apartment buildings, and expansive parks as the taxi drove me north. The road turned wide on the highway as we passed flat, yellowing fields, with the mountains in the rear view. Finally, we reached our exit and wound through an industrial park, each mysterious complex protected by steel gates.

When we arrived, we were met by a series of commercial trucks maneuvering through the driveway. The facility was humming with activity, even at 9:00 a.m. on a Monday of a holiday weekend. The unending blue sky arched overhead as I followed the path through a meticulously manicured lawn, bushy green shrubs, and spiny pink plants crowded against the buildings. As I reached the front door, Scott appeared around the corner, waving hello and welcoming me into the tasting room.

The room was warm and airy, with an expansive feeling created by the wooden floors, used barrels, tall ceilings, and big windows. Tables of various shapes and sizes were strategically placed throughout the room, with a wide tasting counter in the front displaying the distillery's products. The distillery makes twenty-three types of spirits, ranging from whiskey and vodka to gins and liqueurs. All but one of its products are made here. Most of the ingredients are sourced locally, within twenty-five miles of where I was standing.

There are many things that make Leopold Bros. unique—their independence as a small, family-owned business, their commitment to using traditional techniques common before industrialization, their efforts to become a zero-waste facility—but perhaps the most significant is that they are the only commercial distillery in the United States, and one of just six in the world, to malt their own barley. The first iteration of the brothers' malting facility was an empty room in the distillery where they could spread barley over the open cement floor. When I arrived, they were nearing the final stages of construction on their new malting floor, which encompassed an entire building.

Malt—a germinated cereal grain that is dried during the malting process—is the magic ingredient that transforms grain into alcohol. The process goes like this: Grain, which includes malted barley, is first crushed into small pieces. This is called the *mash*. Hot water is added, allowing the mashed grain to steep, along with yeast that is added to the mixture. As the grain mash is hydrated, the malt enzymes are activated. These enzymes convert the starch from the grain into sugar, which the yeast then eats. As the yeast eats the sugar (fermentation), it creates two by-products: carbon dioxide and alcohol. The carbon dioxide burns off, leaving the alcohol. Fermentation usually takes three to five days, and after that, the batch is ready to be distilled.

On its own, this scientific process is unremarkable. But when people and politics are introduced, things become more complicated. Before Prohibition, most distilleries malted their own barley. When 1920 rolled around and alcohol production became (mostly) illegal, the whole

American industry was disrupted. People stopped malting for themselves, and many facilities were dismantled and gone forever; others relocated to Canada.

When distilling became legal again, in 1933, few producers had the ability to malt barley themselves, so they had to outsource to commercial facilities. Then, after World War II, malting became mechanized, and commercial malting plants almost entirely replaced in-house operations in the United States. The same thing happened in Europe, and now only six distilleries in Scotland malt their own barley. Today, just a handful of companies make malt and sell it to breweries and distilleries throughout the United States and Europe.

By the time Todd got into the business, distillery-operated malting floors had long since vanished across the globe. He wanted to bring them back, relying on traditional techniques he had learned in school and while apprenticing in Europe. When the brothers moved to their current home in North Denver in 2014, they became the first US distillery in decades to malt, and the only one with an operational malting floor.

Malting their own barley also aligned with the brothers' environmental mission. The elements that make their operation eco-friendly are relatively straightforward. First, the Leopolds source their barley locally, from within twenty-five miles of the distillery, which cuts down on the energy needed for transportation. Many of the other ingredients they use also grow in Colorado, so they can create completely local products. Second, a portion of their operation runs on solar power, using those powerful rays that burned my skin on the first day I arrived in Denver, rather than fossil fuels. Finally, they employ the system Scott originally created for breweries in order to save water.

Todd had long had his sights set on a new malting facility, something bigger than the malting floor that overlooks the tasting room. Construction is in full swing on the day of my visit, a crane behind us and construction workers in constant motion. As we leave the tasting room, I follow Scott through a series of hallways until we reach a set of thick plastic strips that hang neatly in a wide doorway. He hands me a white hard hat, a neon-orange vest, and a pair of thick, clear goggles.

"This is our new malting facility," he says, with a hint of excitement in his voice. "It's an active construction site, so watch your step."

As I part the plastic strips, I step into a large white room with a high ceiling. A modern steel grain elevator is in front of me, suspended from the second floor, surrounded by steel beams that hold it in the air. White PVC pipes, wooden crates, orange cones, and blue disks are scattered throughout the room, and fluorescent lights hang overhead. Industrial fans are placed strategically where the wall meets the ceiling, helping to regulate the temperature.

When barley is malted, it doesn't look like much is happening. The idea is to trick the plant into thinking it's still in the ground. Todd collects the grain in the center of the floor and slowly spreads it out. He gathers piles about six inches high. The temperature of the room is tightly controlled, about 55 degrees Fahrenheit, which requires strict monitoring because of Denver's dry climate. Todd prefers this temperature, which is more traditional and mimics that of Scotland, whereas most other commercial malters keep their house ten degrees warmer.

The height of the grain piles and the temperature of the room are meant to create an exothermic reaction. When the barley is piled into a heap, the individual grains emit heat. As they warm, they start to germinate, or sprout, which then becomes malt. It's a fine line: the grain needs heat to sprout, but it mustn't overheat. "We fuel the plant's growth, and right when it starts to develop little rootlets, we arrest its development and dry it all back out," Scott explains. He then cools the grain by shuffling through the pile of barley with a wooden shovel, scooping grains and flinging them into the air. They lose heat while airborne before landing softly back into the pile. This whole process, from start to finish, takes about a week.

The Leopolds' malthouse signifies a new stage for the company. It totals 30,000 square feet, enough space for them to malt ten times more barley than they could in their old malt room. Now they are making more than they need themselves, so they are selling malt to local breweries and distilleries.

Malting is a key element of the Leopold Bros. process, but it's just the first step. The brothers also take a distinctive approach to fermentation. Scott leads me to a cluster of wooden tanks, the bready smell of yeast wafting through the air. I peer over one of the open tanks and watch as little bubbles of carbon monoxide burst near the top. I could let time disappear, watching the dancing surface of the mash, but Scott brings me back to the present with a question.

"Do you know why we use wood fermentation tanks instead of steel?"

"So you don't have to use chemicals to clean them?" I ask timidly.

"That's right. You can't get chemicals out of wood at all—period," he explains. "So all we're doing is draining these tanks, pumping the mash over to the stills, and cleaning them with hot water and tails of the alcohol to disinfect them."

"Here's where the process changes," he says. "You'll notice the tanks are open-top. My dad is a landscape architect, and he and Todd planted specific plants outside that generate wild yeast and bacteria. We keep the windows open, and the fan will pull a draft across the tanks and exhaust the air with the yeast and bacteria into the top of the tank. This entire building works like a tool."

As if on cue, a bee flies through the open window behind us and casually buzzes around a tank at the end of the row. Sensing there is no pollen here, it returns to the shrubs outside.

Todd lights up as he explains the process. "From an environmental perspective, all those yeasts and bacteria that are created during fermentation are going to embed themselves in the wood. The next time you introduce a liquid into it, they'll come out of the wood to saturate the batch. They carry a house flavor."

"It's almost like a cast-iron skillet, where you're building up flavor over time," I offer.

"Yeah. And, what's really cool, if my brother's daughter takes over the distillery, however many years from now, she'll be working with the same flavors that her father developed thirty years prior."

The next stage in the malted barley's life is as alcohol. After the mash is done fermenting, the yeast sufficiently full of sugar, it moves to the still and is distilled, resulting in a clear liquid that gets either aged or bottled. Their vodka and gin are good examples of why malting is so significant. The brothers are not only saving resources by making everything themselves, but are giving their products a distinctive terroir, just like wine. They are capturing Colorado in the bottle.

Todd makes one type of vodka, and the still has its very own room. The shiny copper column is tucked into the corner of the distillery, between the pot stills and fermentation tanks. It's sectioned off by floor-to-ceiling glass windows that are at least two stories high. The column is just as tall. My neck aches from peering up at it as Scott explains its nuances.

The column is so high because Todd wanted more plates in the still. "This thing is basically the same as the other spirit stills, it's just on steroids," Scott says, gesturing to its height. "The greater number of plates, the greater degree of separation. That means two things: higher-proof alcohol and less flavor."

Although flavor is something that distillers generally want more of, vodka is an exception to the rule. Since it's used as a base for their gin and all of their liqueurs except the maraschino, a neutral-tasting spirit is preferable because the flavor will come from the ingredients they infuse it with.

While a spirit like whiskey comes off a still at around 130 proof, with the oils and grain flavors present, vodka comes off the still at 190 proof. The oils have been pulled off during its long journey up and down the column, rendering it a blank canvas for Todd to do his magic. Its high proof also, technically, makes it a hazardous substance, so it gets its own room, complete with a different set of electrical wires and a sprinkler system. Todd's vodka has a few different lives. It's bottled and sold on its own, just as it is, called Silver Tree Vodka, and it's used to make Rocky Mountain Peach Liqueur.

Unlike the Leopold Bros., most distilleries buy vodka from a mass producer, such as Midwest Grain Products of Indiana, commonly known as MGP. The company, famous for its lack of transparency, tight-lipped employees, and windowless buildings, makes spirits and sells them to private labels. MGP produces whiskey for more than fifty whiskey brands. It's the spirit industry's prime example of industrialization; ingredients are detached from the food system and treated as a commodity.

With faceless companies like these, it's hard to monitor where the raw material is coming from or what's going into the bottle—and into our bodies. The base spirits must be shipped to the distillery, which takes energy and resources for fuel and hauling. It's a wasteful process, but one that most distillers are willing to use to get a profitable product to market.

Hundreds of liqueurs are currently available for sale in the United States, many of them using vodka (sometimes referred to as a neutral grain spirit) as their base. This is where things get sticky. Most of these liqueurs are fruit-flavored: grapefruit (or *pamplemousse*, if you're fancy), peach, banana, and cherry. While many companies do include fruit in the ingredient list, there are a host of artificial additives in the bottle, too. Like high-fructose corn syrup. Like red dye no. 40. Like caramel coloring. Like petroleum. Artificial sweeteners, thickening agents, and synthetic chemicals. Things we don't want in our food, but tolerate in our drinks because we're likely unaware that they are there.

Not the Leopolds. Their rule of thumb is to treat alcohol like food. "If I wouldn't eat it, why would I drink?" Scott emphasizes. Instead, they visit farms, talk to growers, and expect crops to be slightly differ-ent each year. They adjust accordingly, with the seasons. This makes everything taste better. It's fresher, more vibrant, more alive.

Their vodka has another identity, too: gin. "All gin is basically fla-vored vodka," says Scott, "so that's the base." Gin, by definition, is a neutral grain spirit that must have juniper in it. That's it. It can have any other type of botanicals, as long as juniper is one of them.

Todd practices fractional distillation when making their gin. This means that they distill each botanical individually and blend the batches together. Why do this? Botanicals, like food, don't all cook at the same temperature or for the same amount of time.

"What Todd did, which speaks a little bit to the black magic of dis-tilling, is recognize that every botanical is going to cook differently.

They all have different boiling points," Scott reasons. As he's telling me this, I notice big plastic jugs, called carboys, that are clustered near one of the stills.

"You can think of it in terms of food," Scott continues. "You wouldn't take a steak, potato, and a vegetable and put them all in the oven at the same time and expect a flawless meal. Todd separates everything, so he's making his decisions with individual plates in each still and cuts on one botanical. So he's just distilling juniper. And then coriander. Then bergamot. He's getting the hearts of every single botanical—only the purest flavors and purest alcohols—and blending them together." The carboys contain each of these batches, waiting for Todd to blend them.

Scott tells me that juniper is one of the very first botanicals to distill off. That's where astringency comes from, that piney taste that reminds me of the cleaning spray my mom used to dust our house. When all of those botanicals are loaded up in the still at the same time, the flavors get muddled, losing their distinct profiles. "It's easier, cheaper, and faster to distill everything at once, but the trade-off is that while you are waiting for the hearts of citrus, which come off later, by design you're capturing the tails of juniper, the heads of coriander, the hearts of coriander, the tails of coriander, the heads of the next thing, and so on," Scott explains.

He gives me one more example to really drive his point home. Some time ago, the History Channel did a show on spirit production, which Scott watched. It featured one of the largest gin producers in Europe. The head distiller was stationed in front of a still, describing his process. He said that he could be blindfolded and thrown into the distillery and he'd be able to tell exactly what time it was because of what he was smelling.

"He's telling you that if he smells juniper," Scott says, "he knows it's 10:00 a.m. because it's the first thing to come off the still. If he smells orange peel, he knows it's 1:00 p.m."

Though fractional distillation may take longer and be more labor-intensive, it makes malting their barley in-house and using it to produce vodka worthwhile. In the end, they are using fewer resources, taking up a smaller ecological footprint because multiple ingredients don't need to be delivered, and creating a gin that has a crisp, distinct flavor, the kind that really makes a Negroni shine.

Distilling their spirits this way gives their products a different shelf life than most. Because no artificial ingredients are added, their spirits aren't meant to last forever, or taste exactly the same each year, just like food. And also, like food, or wine, or beer, when a bottle is opened, it will slowly oxidize as it interacts with air. What's in the bottle might taste slightly duller four or five or six months down the road.

But this is also where fractional distillation helps. Ingredients will likely taste a bit different each year, depending on weather and growing conditions. By distilling everything separately and then blending it together, Todd can work within the flavor framework that gives the Leopold Bros. products their identity, making them taste similar each year. Consistency over time is hard to promise (though many companies do, using additives and commodity crops to deliver); blending is an alternative that doesn't compromise the Leopolds' values. It allows them to make the most flavorful gin or liqueur they can.

Todd, who has been running around in the background in his signature duck-brown Carhartt overalls, gently interrupts our tour.

"Sorry, Scott, but we need you for a meeting," he says, flashing me an apologetic smile. The brothers are off to discuss the malthouse, which is scheduled to open in just a few days, the hum of the construction still in the air.

Scott escorts me back to the tasting room before heading off to the meeting. As I'm waiting for my taxi to take me to the airport, their father, Bob, wanders in. He introduces himself, his curiosity about who I am palpable. I tell him I'm there to hear about their sustainability work, and he tells me about his background as a landscape architect, pointing to different plants through the window, not missing a beat. He speaks about each of his sons with pride, his eagerness to be involved with the family business evident in every sentence.

My car arrives at the same time that his wife, Joan, pokes her head in to grab him for their own meeting. We excuse ourselves, Bob making his way back into the distillery, and me making my way out, cranes skating behind me as I drive away into the sun.

CHAPTER 4

Rum

Every time I'm dining at a restaurant that promotes itself as farm-to-table, the kind that lists its purveyors and the farms where its ingredients are grown, I ask to see the spirits menu. Or, if I'm sitting at the bar, I'll spend a few minutes scanning the bottles that neatly line the shelves behind the bartender.

"What bitter liqueur do you use to make your Negroni?" I'll ask. What I really want to know is whether the bar uses Campari, a bold Italian aperitif known for its bright red color and viscosity. Campari is so common that it's become synonymous with a Negroni. You rarely see a substitute, but, nevertheless, I always ask.

Usually one of three things happens next: If a substitute is available, I'll order my cocktail with that. If my only choice is Campari, I'll order a different drink. The third scenario, which can be tiresome for my dining companion, especially if it's John (since he's heard this routine countless times), is when the bartender will cock her head and ask me why I'm asking, wondering what I have against Campari.

It's not that I don't like the taste of Campari. Quite the opposite, actually. Its taste and texture are so iconic that its flavor profile is the baseline for other aperitifs. If an alternative is much sweeter, people think it won't balance a Negroni. If it's thinner, people think it will get lost in the drink, despite its bitterness.

My issue with the liquor stems from that eye-catching red color, which doesn't naturally occur during the distillation process. Campari, like other companies that make aperitifs, used to dye their liqueurs with a cochineal bug (sorry, vegans). These beetles, which are native to the Americas, imbue liquid with the color carmine. It was a natural dye, nothing that was made in a lab by mixing synthetic chemicals. That changed in 2006 when the company cited uncertainty about the longevity of the supply of cochineal bugs. Campari switched to red dye no. 40, an artificial coloring agent that has been linked to cancer, stomach issues, migraines, anxiety, and allergic reactions. It's found in many common foods, like Skittles, Jell-O, and Gatorade.

If people are aware that they're consuming red dye no. 40, that's their choice. But if you're a restaurant or bar owner who claims to carefully source ingredients that are organic, non-GMO, and grown without help from pesticides, carrying alcohol with synthetics completely contradicts your other efforts. Simply put, the same principles are not applied evenly to both food and beverage. Wine and beer may get some attention, but spirits usually fall by the wayside.

That's why I insist on annoying bartenders with my questions—because we don't ask enough of spirits. We don't expect as much. But cocktails are ingested the same way as our meals, a sip and a swallow, and deserve to be more than an afterthought. It may not occur to people who think of themselves as mindful eaters, aware of the food system, to question why a spirit is bright red, but it should. We have the power to spend our money on alternatives, to ask what else is out there, to customize our Negroni orders.

The issue goes beyond just red dye no. 40. In the United States, alcohol production and particularly the labeling on bottles are regulated by the Alcohol and Tobacco Tax and Trade Bureau (TTB), not the Food and Drug Administration. Though the FDA has a list of ingredients that are generally recognized as safe (that is, GRAS), which applies to both food and spirits, the government doesn't regulate spirits the same way as food. The TTB regulates and collects taxes on trade and the import of alcohol (and firearms and tobacco), which essentially deals with the business side of the industry, leaving the production— and everything that goes into it, like farming, distilling, and bottling— out. The regulation is about taxes and trade, not about what can and cannot go into a spirit. The FDA, on the other hand, is responsible for protecting public health by ensuring the safety, efficacy, and security of food (and other things we put into our bodies, like drugs and vaccines). If the federal government doesn't see the importance of treating alcohol like food, or of transparency from those who make it, the general public won't either, furthering the divide between public perceptions of food and alcohol.

That GRAS list includes many ingredients used by industrial distillers, including coloring agents like red dye no. 40; sweeteners such as high-fructose corn syrup; tannic and polyphenolic substances, which are insoluble in water; inorganic salts; essential oils derived from herbs and spices; flavoring agents, like caramel; and petroleum, which adds body. Many of these things are made from genetically modified crops.

These additives, most of which are artificial, often take the place of more natural, recognizable ingredients. That strawberry liqueur in your favorite summer cocktail? There are probably no real strawberries in it. The flavor was created in a lab, added to the base spirit after it was distilled. The blueberry vodka you drink with soda? You're better off muddling fresh blueberries in yourself. That cinnamon flavor in the whiskey you used to drink when you were in your twenties? It came from a flavoring agent. The herb-flavored *crème* that gives your cocktail a green color? That hue came from a dye.

Another trick that many producers use is to create the illusion of age by adding caramel. After a spirit has matured in a barrel for a period of time, it comes out brown, the barrel giving it both flavor and color. Often, the spirit's proof is quite high and needs to be watered down to a drinkable ABV—so it's not too hot or unsafe to swallow—and this process lightens the color. To correct the color, distillers and blenders might add caramel to darken it. This is a practice so common that it's controversial to take a stand against it. But many companies aren't transparent about what they are adding, whether it's caramel, honey, or something artificial, or disclosing where those ingredients come from. And this is the problem.

As someone who values transparency from producers, I like them to tell me up front that they don't add anything to their juice. And, if they do, I like them to tell me that, too, so I can make an informed choice about whether to drink it. This is what led me to Equiano, an Afro-Caribbean product that was launched in 2020. The brand has transparency and traceability, and it uses no additives or artificial ingredients.

～

Equiano is a collaboration headed by Ian Burrell, a widely respected rum advocate and educator. He's responsible for partnering distilleries from two countries—Gray's Distillery in Mauritius and Foursquare Distillery in Barbados—to create the world's first Afro-Caribbean rum. While each country that produces rum typically has its own unique style, it felt natural for Ian, who identifies as African-Caribbean, to blend the two cultures together. A Jamaican who was raised in England, he lives that life every day.

From the start, it was important for Ian to partner with Richard Seale. Richard is a fifth-generation blender from Barbados and the first in his family to distill. He's what many call a "purist," and he doesn't think that rum should have any additives, including sweeteners or coloring agents.

In a 2019 interview for *Forbes* magazine, Seale had this to say about the topic: "My mantra is 'Drink what you like, but know what you are drinking.' I make no apology for this. If you are drinking inexpensive spirits, it matters little what you know about the product you enjoy, but the moment you enter the rarefied world of paying premium for spirits, not only should you know, but also your appreciation and enjoyment can increase immensely."

Ian feels similarly. "I always say, 'Drink what you like, enjoy what you like, but know what you're drinking,'" he tells me over Zoom.

I ask why it is important to him that his rum not contain any additives.

"It's a great question, but controversial in rums," Ian explains. "The reason why it's tricky is because, historically, rums were made the same way as a lot of whiskies—it was all about how they were aged and the flavor that came from fermentation, how you distilled, and the barrels you used. All of a sudden, people started adding stuff to make them either look older or taste sweeter or smoother. It was a marketing ploy to get more money out of a particular product."

Ian tells me that additives have become ingrained in the way that various regions make their styles of rum.

"Certain practices in certain areas are historical," he says. "In places like Guatemala and Panama, they've been adding sweeteners for many, many years. But that is alien to someone who has been making rum in Jamaica or Barbados or St. Lucia."

Someone like Richard Seale. Richard, who founded the Guardians of Rum, a group whose goal is to protect the cultural identity, authentic practices, and geographic influences of traditional rum producers around the world, asserts that sweetened rums are often mass-produced. But artisanal rums, like what he makes with Foursquare, do not need sweeteners because they only mask the quality of the alcohol.

"This is why Richard is Richard Seale," Ian says. "He is 100 percent against additives. He'll call people out for using them."

And call people out he does, mostly on the internet, which has earned him some unflattering nicknames. But it's this zeal, this commitment to making a product representative of Barbados's history, culture, and agriculture, that has given him a reputation as one of the best rum producers in the world. It's also what drew Ian to him in the first place, crediting Richard with taking rum to a whole new level. Foursquare is one of the most coveted spirits in any category because only a limited amount is made each year. It sits on the top shelf with a few other exclusive spirits that can cost over $100 a bottle. Foursquare sells for thousands of dollars on the secondary resale market, much like the coveted Pappy Van Winkle bourbon from the Buffalo Trace distillery.

"Richard is an important part of rum's story, especially in modern history—everything he does around it and how he's planning for the future," Ian says.

The respect is mutual. Richard credits current interest in the category—from both consumers and the media—to Ian, who organized the world's first Rum Fest in 2007, after working as both a bartender and a brand ambassador. The festival has since expanded to various cities around the world.

"Rum in London at a time was just dark, gold, and white," he says. "That was it. No one asked where it was from. You couldn't really talk about different origins like Jamaican rums, Barbados rums, agricole. No one knew what an agricole was back then, unless you lived in France. It was just very basic. Rum definitely was one of the last spirits to become premium and for people to discover what the differences were between the different types of rums."

Ian spent his own money to bring rum producers from around the world together in London to speak about their products and lead guided tastings in order to educate people about the complexities of the spirit.

And complex it is. Rum is an integral part of colonial history, its past dark and violent. The spirit is made from sugarcane—either fresh cane juice or molasses—which grows best in tropical areas. Sugarcane was an essential crop during colonization in the Caribbean and on the islands off the coast of Africa—Madeira (where Christopher Columbus's father-in-law owned a sugar plantation), the Canary Islands, and the Azores. Barbados, a tiny Caribbean country just twenty-one miles long and fourteen miles wide, became one of the world's largest producers. Introduced by the Dutch in 1640, sugarcane grew easily there, its fibrous stalks absorbing the water and growing tall under the hot sun.

Clearing land to plant cane was grueling work and fields were burned to harvest it—a practice that persists today—leaving enslaved workers to pull the stalks by hand as the earth still smoldered. Smoke filled their lungs as they navigated the snakes, rats, and scorpions attracted by the cane. The real danger to enslaved people came, of course, from those enslaving them, in the form of beatings, starvation, and inhumane living quarters. It's estimated that thirty-five of every thousand of those who were enslaved in Jamaica died between 1829 and 1832. The average infant mortality rate was 50 percent. Because enslaved people and their babies kept dying, more and more people were captured to work the plantations, ensuring that the colonizers became world powers. The story of sugarcane, colonization, slavery, and the long-term effects of that brutal history is an essential one, but I am not the person to tell it.

Here's the story I am able to tell:

Rum is a staple in Barbados and the Caribbean. Many Barbadians claim that they invented it (as do others). After sugarcane was first planted there in the seventeenth century, people began to distill sugar into "Kill-Devil," the beverage from which rum was later derived. In the nineteenth century, more than ten sugar factories produced the sugar to make rum, eventually giving rise to distilleries like Mount Gay, St. Nicholas Abbey, West Indies Rum, Doorly's, and Foursquare Rum Factory and Heritage Park. Rum has its own style, often described as well balanced and expertly aged. It's usually made from blends distilled from both pot and column stills. According to Richard Seale, "until about the 1680s, Barbados's production of sugar and rum vastly exceeded the other islands." During the nineteenth century, the rum that was made there was mostly drunk there, not shipped out to other countries. But later, distilleries in other countries began to outpace those of Barbados, making more and exporting it to other countries. Producers in Barbados lost money, and distilleries shuttered. Sugar became a commodity crop in other places, too, pushing down the global price. Unfair trading practices took over, undercutting the value of cane further, and sugar beets were planted as a substitute.

"The land itself couldn't sustain growing sugarcane," Ian says, "and then, more importantly, the price of sugar started going down, especially as sugar beets started becoming popular in Europe. There wasn't a market for the sugar anymore."

It's from this history that Equiano was born, and got its name. The brand is named for Olaudah Equiano, an African born in 1745 in what is now southern Nigeria. As a child, he was stolen from his home and sold into slavery to an officer of the British Royal Navy. He crossed the Atlantic on a slave ship, with 244 other bodies packed tightly against his, and ended up in Barbados. Later, he was sent to the colony of Virginia, where he was bought by Lieutenant Michael Henry Pascal, who renamed him Gustavus Vassa—not the first time he was given a new name. Olaudah was brought to England with the lieutenant, where he accompanied him during the Seven Years' War with France. Equiano traded rum on the side and was able to save enough to buy his freedom. He taught himself to read and write, authored a memoir that became a best seller across Europe in 1792, and got involved with the abolitionist movement advocating for the freedom of enslaved people.

"I knew Olaudah Equiano's story because I'd been told about him when I was younger," Ian says. "When we were playing around with names for our rums, I had a lightbulb moment. It made sense. Africa. The Caribbean. He went to Barbados. This is where our rum is. And then he came to London, which is where he learned to speak English, wrote his book, and became the man he ended up being. Our rum makes the same journey as he did in the eighteenth century. Everything just fit into place."

The blend of these cultures allows the Equiano team to use the natural resources of both Mauritius and Barbados, including sugarcane. All of the sugarcane that's used at Gray's Distillery is grown right on Mauritius.

"The nickname for Mauritius is 'The Green Island' because pretty much 99.9 percent of every blade of grass there is sugarcane," Ian says. (This might be an exaggeration, but not by much.) "There is sugarcane everywhere. You can be walking down the street and see it growing out of the walls. There's mountains of it."

In fact, he tells me, all demerara—a style of sugar that is used to make rum (and other things)—that is imported to Europe comes from Mauritius. In addition to their own sugar refinery and distillery, Gray's Distillery owns their estate, so they grow all of the cane that is used to make the molasses for their rum.

Barbados, however, has been subject to a different colonial history, one that left the land so stripped of nutrients that it could no longer support cane crops, giving rise to the sugar beet. The country was forced to import their molasses, like a lot of other islands in the Caribbean that fell victim to the same forces.

"A big tanker comes in, like an oil tanker but with molasses, and stops at various islands. Then the molasses would be sent to the actual distilleries," Ian explains. Most of the sugar used for rum in Barbados comes from other parts of the world, shipped over by boat. Other crops—cash crops that can fetch a higher price—are now farmed on the land.

But recently, Richard has started to grow his own cane. He's hoping to create an estate rum, which means that the ingredients are grown on the company's land. In the meantime, he has relationships with a couple of sugar farms in the northern part of the island, where he buys cane directly. He knows who is growing it and where it's coming from, and he's able to visit anytime, like many of the other distillers who prioritize the environment and transparency.

This traceability is important, especially when looking at the environmental effects of rum. This is where Karen Hoskin comes in.

～

Karen Hoskin founded Montanya Distillers in Silverton, Colorado, in 2008. Before venturing into rum, she had worked in public health for ten years (she has a master's degree in epidemiology) and in design for another decade. When she decided on this career change, she learned everything she could about the spirit, from its history to distillation techniques. She had a few mentors in Colorado who were always available to answer her questions, including Jake Norris, who then worked from Stranahan's distillery, and Dave Thibodeau and Rory Donovan from Peach Street Distilling. Though none of them made rum and weren't using the same base ingredients, they helped her learn about distilling and understand fermentation. She bought a copper pot still and got to work on her alchemy, getting her distillery off the ground.

Sustainability was essential to the company, even in Montanya's early days. Karen views environmental health as a part of public health, and she wanted to infuse an eco-conscious ethos into the DNA of the distillery. She started on a personal level, refusing to buy single-use convenience plastics. It snowballed from there, at home and at work.

A key element was responsibly sourcing her sugarcane, especially since Colorado is not known for growing tropical crops, their tall stalks not meant to withstand the cold and snow of Rocky Mountain winters. Because Karen never expected to source locally, she wanted to find sugarcane that was not genetically modified or harvested in a way that would be detrimental to the environment. Originally, she bought ingredients from Hawaii, but the farms stopped certifying that it was produced in the United States.

"The cane was being pulled from the commodity market in California from HC&S, a company in Hawaii, but when there is a shortage there they mix in a bunch of sugarcane from Mexico or India or wherever," she says. "I realized that I wasn't sure anymore that we had been making our rum from American cane."

It was a reasonable concern. Sugarcane grows in more than ninety countries, with Brazil, responsible for 40 percent of the global total, leading the charge. Cane crops cover 65 million acres of land worldwide, rendering it one of the biggest commodity crops on the planet. In 2017, 1.8 billion tonnes were produced across the globe.

Sugarcane is a type of jointed, fibrous perennial grass. It belongs to the genus saccharum and is in the Poaceae family, which also includes corn, wheat, rice, and sorghum, all of which are also used to make alcohol. Sugarcane can grow as tall as six to twenty feet and takes up to twelve months to mature. It thrives in tropical climates, and it needs a lot of water to grow. The World Wildlife Federation calls it "one of the world's thirstiest crops." It takes 213 gallons of water to produce a single pound of refined sugarcane. That's just about nine gallons per teaspoon.

Growing cane creates a significant environmental footprint, arguably larger than that of any other commodity crop. This is partly because the process is so water-intensive, but also because it's highly polluting. Since sugar is a staple in our collective diets, there's always a demand.

Big swaths of forests are cleared to make room for cane crops, especially in Brazil. This means there's less vegetation to buffer runoff, which pollutes waterways with silt and fertilizers. (Sugarcane refineries also clog freshwater ecosystems with chemical sludge as it's discharged from the mills.) In turn, runoff and deforestation both accelerate erosion and contribute to flooding.

Much as we saw in the case of agave, clear-cutting land to grow cane means the loss of other plants, bugs, and animals that make an ecosystem healthy. There's nothing to slow the spread of plant disease and no natural pest control. Monoculture requires that the jointed stalks be sprayed with pesticides, which will later be washed into streams and lakes. It's a vicious circle, especially when it's happening all over the planet.

But it's more than that. The ecological toll also comes from the way sugarcane is harvested. This can be done in a couple of different ways, one more environmentally friendly than the other.

The first method is the one that is most common, a tradition dating back hundreds of years: burning the fields. Sugarcane's fibrous stalks make it difficult to remove the stiff leaves that protect it and the tough weeds that grow around the plants. Setting fire to the field helps remove the leaves, freeing the stalks of the material that helped them grow, and fire flushes out snakes and vermin. As with all fires, cane field fires release particles and hazardous gas into the air, which is harmful not only to the nearby wildlife and other crops, but also to the people doing the harvesting. Though the plant stores a lot of carbon dioxide, burning the fields releases just as much into the atmosphere. It also degrades the soil, stripping it of the nutrients it needs to support the next planting, and increases the risk of erosion and runoff.

The second method is done with the help of heavy equipment. The stalks are pulled out of the ground by a machine that looks like a cross between a backhoe and a forklift. After the stalks are uprooted, the leaves are separated and left on the field. This helps to prevent erosion, control weeds, and conserve soil moisture, which can improve topsoil health with more water and nutrients for the next crop that will be planted. It's also possible to let a field lie fallow or to plant legumes, giving the land a year off from supporting cane. This method also reduces human labor and makes working conditions safer.

Responsible harvesting, growing, and processing were all important to Karen. While she was trying to look for another source of sugarcane, a couple walked into Montanya's tasting room, which is now in Crested Butte. (Karen relocated the distillery there shortly after she got off the ground because she felt that, in the long run, Crested Butte would be more business-friendly than Silverton.) As it turned out, the couple ran a cane farm in Belle Rose, Louisiana, that had been in their family for generations, and they also had personal connections to Colorado.

It wasn't just their story that Karen liked, it was also the way they handled their crops. The farm, Lula Sugar Mill, is a co-op run by seven families whom Karen now knows personally. Before she decided to buy directly from them, she went to visit the farm in Louisiana.

The Lula Company has three components: the growing co-op, which consists of the seven families who grow the sugarcane; Lula Westfield, the mills that process the cane; and the refinery that turns it into the form of cane that you and I buy at the store. Karen interacts with only the farm and the mill, because the cane is sent to her before it is refined. When she went to visit, her priorities were to see how they were growing cane, how the mill was handling waste, and how the entire company was treating their workers.

Sugarcane harvest season is from August through the end of December. "Luckily, when I went down, it was harvest time," Karen says. "I was able to see every step of the process in person, from them harvesting the cane, bringing it to the mill, processing it at the mill, all the way to seeing how the mill was generating all of their electricity and heat, how they were treating their wastewater, and what they were doing about emissions with their stacks in the mill. Everything."

Karen was impressed with what she saw. The cane was harvested by machine. The workers sat in air-conditioned cabs while the equipment pulled the sugarcane from the ground. This means that none of the fields are burned and none of the cane is harvested by human hands, which, in this case, is a sign of social and environmental responsibility.

Working conditions were a particular concern for Karen. Around the time she visited Belle Rose, there was news coming out of Nicaragua that the rum distillery Flor de Caña was grossly mistreating their workers, who were forced to harvest cane by hand, burn fields, were not given adequate access to water or shade, and were compelled to work incredibly long days because they were being paid by the pound. The mortality rate for sugarcane harvesters there was over 20 percent higher than for other occupations. But at Lula, the harvesters are treated well and paid a fair wage, and they receive benefits.

Lula's owners, Isabel and Will Thibaut, describe just how labor-intensive growing and harvesting cane can be, even under the best working conditions: "Planting begins around Labor Day, using a portion of last year's crop: the tops of existing stalks are cut and then laid in the ground. Much like bamboo, eyes at each joint grow into new stalks. Once the cane is planted, harvest of the previous year's crop begins, typically around October 1. Weather depending, harvest runs 24/7 and ends around Christmas. A wet year like 2018, however, can push that into the first or second week of January. Then the harvest has to be transported to the mill."

Lula can process 12,000 tons of cane in twenty-four hours, employing eighty-five workers during grinding season. The company website is full of information about their production methods, their processes, the types of machines they use, how they wash and prep cane, how they grind and clarify and heat, and the specific number of blades each of their mills uses, complete with the make and model. If you want the details, Lula has them for you. If you want transparency, that's what you've got.

I asked Karen how the farm addresses soil health, given that sugarcane is almost always a monoculture crop. "You can't do a periodic crop rotation for cane to restore natural resources, like nitrogen to the soil, because it's really the only thing that grows well in the bottom land," she says.

Instead, what happens is that bottom land gets floodwater from Lake Pontchartrain on a regular basis, which Karen says is essentially like crop rotation. It brings in high-nitrogen soil from the bottom of the lake during storm cycles, which happen every year, to re-nitrify the land.

Over at the Lula Westfield mill, processing the cane uses a lot of resources, including heat, energy, and water. To minimize the impact, the mill is entirely biomass-operated, meaning that they create all of their electricity and heat from the bagasse, or the fibrous solids of the cane. Though generating energy this way isn't entirely unknown in sugarcane milling, it is becoming rarer as the supply chain changes and mills get farther from their sources.

Karen also cared about air and water quality. Because she is thorough, she recruited the University of Louisiana to go through a certification process to assess the air and water on its way out of the mill. They found that because of Lula Westfield's treatments, both air and water were actually cleaner when they left the mill than when they went in.

"The cane fields are right next to the workers' own houses, right in their own communities, so it's essential they take good care," Karen says. "That's the long and short of it."

Once the cane is milled, it is heated in a boiler and then put into one-tonne pressurized vacuum tanks called *supersacks*. There's a little more to the process, but it's highly technical. In fact, it's so specialized that people from Honduras have to travel to the United States each year to do this work because they are among the few in world who can claim the necessary skill set. After they perform their magic, the cane gets shipped by freight truck to Crested Butte, Colorado. The cane isn't refined, so it has a lot of impurities, like rocks, left in the sack. But skipping the refining process saves energy and means that the cane isn't exposed to sulfur dioxide and other chemicals such as calcium hydroxide and phosphoric acid. Once the cane arrives at the distillery, Karen and her team start the distillation process.

Sharing all these details is part of Karen's effort to be radically transparent. She isn't hiding anything, including the touch of Colorado-made honey that Montanya adds to color some of their rums, which they say is less than 0.4 percent per bottle and helps bring out flavors in the alcohol.

Montanya became a B Corp in 2018 after a lengthy certification process that evaluated the company on four different criteria: social, environmental, community, and governance practices like equitable hiring. Montanya qualifies as a manufacturer, so they had to measure everything they were doing, down to how much heat they use and their volume of compost. In the end, the distiller didn't have to change any of their practices, they just needed to communicate their social and environmental efforts. In a 2018 interview with Think Radio, Karen said, "I knew we were doing it, but now we have to measure it."

Montanya does a lot more than just source its cane from a family-owned co-op that doesn't burn their fields or exploit their workers. Montanya's tasting room is powered solely by wind; they offset their carbon production and post their stats online; they use recycled heat produced by their stills—which cook over an open flame—to heat their facility; they don't use chillers on their fermentation tanks; they ship in bulk when possible; they don't use plastic packaging; they print with soy-based ink; they use passive solar energy in their barrel room to regulate the temperature while the rum ages; and the list goes on. And on.

What Montanya is doing isn't just new for rum—it's new for a lot of distilleries. There are only three other distilleries with B Corp status in the United States, and just five internationally. If Karen had her way, the Alcohol and Tobacco Tax and Trade Bureau (TTB) would ask twelve questions of all rum producers in order to promote transparency. She advocates for disclosing: where the cane is sourced; the type of still that is used to make the rum; what, if any, additives are used; how long it was aged; if it was aged in anything else in addition to the barrel; if the rum is blended; if the solera method was used, wherein some of the liquid from the barrel is added to a new batch so old and new are constantly being blended; and where the water to proof the rum—or bring it down to a drinkable ABV—comes from. Though there's no shared international understanding of what environmental and social responsibility looks like for rum, asking a distillery to be open about all of these questions is a step in that direction.

Karen is a busy woman, running the distillery and giving talks around the country. She often talks about environmental issues, pushing people to engage with the little things in order to solve the big, systemic things, together as an industry.

Though rum is a nuanced category with a tangled past, people like Ian Burrell and Karen Hoskin are at the forefront of unraveling its layers so that all can enjoy this spirit. Whether it's honoring the people and places that have shaped rum, or advocating for a regenerative way of making it, Ian and Karen share a commitment to radical transparency. They want us to know what we are drinking so that we can make informed choices, recognizing rum's connection to our history and our future.

CHAPTER 5
Brandy

Fall in the Bay Area is a magical time of year. The dense fog that hovers over the steep hills starts to lift, making room for blue skies. The days are warm and sunny. It's the closest we get to summer weather here, even as the rising temperatures usher in harvest season.

At St. George Spirits, fall means it is time to make pear brandy, an unaged *eau de vie*. Built in a former naval airplane hangar, the distillery sits on the eastern shore of Alameda, just across the bay from San Francisco. From August to October, freight trucks pull into the loading dock, their flatbeds full of freshly picked pears. About half a million pounds make their way to St. George each season, winding through a tunnel, or over a bridge, from Oakland toward the water.

With the San Francisco skyline, the Bay Bridge, and the bay itself in the background, the beauty here is distracting. The slight breeze doesn't do much to alleviate the fall heat, so it's usually warm when you step inside the distillery. The staff maneuvers around a smattering of waist-high boxes arranged like puzzle pieces, green fruit brimming over the top. It smells like an orchard, with the fragrance of the ripe pears wafting through the air.

St. George was founded in 1982 by Jörg Rupf, who immigrated to Berkeley, California, in the 1970s. He grew up in the Black Forest, near Alsace, where Germany borders France. In his village, everyone made brandy or beer or both in their backyards—his family included.

"It was just part of the local culture," he says. "We had a plum tree in our yard, and my dad put me in charge of making spirits from that," Jörg says. "We didn't have a still at home, but there was a local distiller just fifty yards away from our house. I harvested plums every week."

Jörg learned to follow a farm-to-table philosophy before it had a name. When he founded St. George Spirits in 1982, he approached spirits like a vintner does wine, concerned about capturing the essence of fruit in the bottle. He worked with family farms and local purveyors to source his ingredients. The business was (and still is) independently owned, making it the first craft distillery to open in the United States since Prohibition ended in 1933.

Jörg believes that you should be able to taste and smell brandy's raw ingredients without the interference of artificial additives. He passed that philosophy on to Lance Winters, his apprentice, who became sole proprietor after Jörg retired in 2010. Jörg imparted it to Dave Smith, too, when he joined the team in 2005. (Dave later became a partner in the business, in 2014.)

Almost forty years after Jörg started St. George, his products are still made this way. The difference is that now local fruit is less available, farming methods are more inconsistent, fire has ravaged farmland in Northern California, and droughts are threatening to dry up reliable sources of water. So it's more complicated now than when Jörg started out, with many more environmental pressures.

On the day that I make my way over the Bay Bridge to visit St. George, as I've done many times before, the pears have arrived and the distillery is humming with activity. Stills are running, fruit is fermenting, and tourists are visiting. James Lee, a tall, mustached employee in overalls, is driving an orange forklift, its long steel forks lifting boxes of pears. The pears will soon be crushed, their skin, core, and juices mashed into an aromatic pulp. Lee picks up a pear between trips on the forklift and bites into its flesh. This is something he does over and over again each harvest, monitoring the ripeness of the fruit. "It's one of the perks of the job, getting to eat all of these pears," says Lance.

On the far side of the hangar is the bottling line. When it's time, bottles move along a short conveyor belt, filled, corked, and labeled along the way, the occasional clink of glass echoing through the room. It's here that all of the work—in the fields, in the stills, and in the bottles—comes together. It's in these bottles that eager drinkers become privy to the bounty of the year's harvest, to the terroir where the pears were grown. From the farm fields that surround Northern California, St. George has cultivated a unique flavor profile, one that is distinctly its own.

The sights, sounds, and smells in this building have marked fall for decades. So how does St. George maintain that flavor profile year after year, when pears change every season? How do all the elements beyond the company's control—fire, drought, climate change, land management, and access to pears—impact its brandy, which is their flagship spirit? This is what I'm here to find out.

"Can you tell me about where you are getting your pears from and how you decided to buy pears there?" I ask the two bald men who are sitting in front of me.

The pair are Lance Winters, who is fit, tall, and in his fifties, and Dave Smith, who has a similar build but is a little shorter and several years younger. I've known them both for a few years now, ever since I started interviewing Jörg. I'd show up, my camera gear slung heavy on my shoulder, to ask Jörg hours of questions, documenting his oral history. There are a lot of people—writers and otherwise—who have visited the distillery over the years to ask the team about their work, but I never stopped coming, long after my initial interview sessions with Jörg ended. No matter how busy Lance and Dave are, running the distillery, they have always graciously agreed to answer my questions, which are often about industry trends, best practices, or environmental issues—as I'm here to discuss today.

In Lance's office, I settle into an old overstuffed leather chair, my back to the window that looks across the bay toward San Francisco. Both men wear T-shirts and faded jeans, the unofficial St. George uniform. Lance chews on his unlit cigar as Dave swirls clear liquid around in a glass bottle, and they begin to tell me about how they source their pears.

"For the longest time, we were getting all of our pears exclusively from California," Lance says. "They were grown in Lake County and Mendocino, and we would buy them through the Thomas Brothers packing shed."

Jörg began with pear brandy back in the early 1980s, when he found himself in a wooden shack in the back of a small winery on the outskirts of Oakland—with a box full of Mendocino pears. In his early days, he spent a lot of time sourcing the perfect pear.

"I would go up to Napa Valley and Brentwood to get fruit," he told me. "The farmer didn't sell it, so I picked it myself." He found pears in Mendocino that reminded him of those from his youth in Germany.

These pears, sourced locally and picked by hand, became the foundation of St. George. This is how he built his relationship with Thomas Brothers.

Each year, the Thomas Brothers family would send Jörg samples from different orchards. Each year, the pears would be crushed and distilled in individual batches. Each year, he would decide which batch had the flavors and characteristics he was looking for. Then he would choose which local orchard, just miles north of the distillery, he was going to buy from. It went on like this for years.

And then the local supply of fruit became unreliable. Agricultural land in Northern California grows more expensive by the year, the value rising continually since 1994. In 2012, irrigated cropland cost $12,000 per acre, and nonirrigated cropland cost $3,550 per acre. Farmers want to plant crops they know will be lucrative enough to compensate for those high costs, and in Northern California, wine grapes yield the highest prices.

"Over the years, we've seen more and more California pear orchards get ripped up to make way for pinot grapes," Lance says.

While I live just an hour south of Napa and Sonoma counties, I hear more about farming in the Central Valley, where nut trees dominate agriculture, soaking up precious water resources from their northern neighbors. Wine is all around me, but I had no idea that it affected pears.

"Pinot makes more money per acre than pears," Lance continues. "It's easier to pick and easier to sell. Farmers tend to go for that." As farmers cleared out their trees and planted grape vines, St. George has had fewer and fewer sources of California-grown fruit over the last twenty years.

"It's just getting more and more dried up, specifically since pinot started to take off in the early 2000s," Dave adds.

Aside from farmers choosing grapevines over pears trees, California pear growers face other threats. Fire and drought have defined the state, up and down the coast, ever since cities sprang up in the 1800s, and lately climate change has intensified matters. The last few years have produced devastating fire seasons, burning hundreds of acres of farmland and residential property alike.

In November 2018, the Camp Fire erupted, burning the town of Paradise to the ground in a matter of hours. At least eighty-five people died. More than 18,000 structures were enveloped in flames. This was California's deadliest and most destructive fire on record. I remember eating my lunch outside that day on UC Berkeley's campus and smelling the smoke from 160 miles away, hours before I heard the news. But it wasn't the only blaze to break out; there were more than 8,527 wildfires across the state that year.

In October 2019, the Kincade Fire started in Sonoma County, tearing through 77,758 acres and becoming the largest fire on record. Five people died. Pacific Gas and Electric (PG&E) pre-emptively shut off power to thousands of residents all over Northern California, leaving many without access to emergency service warnings for evacuations. A friend slept on my couch in San Francisco for two weeks, afraid to return to her home in Napa, while my colleagues from Sonoma County fled, locking their doors behind them and hoping they would have something to come back to later. Where I live, fire feels inescapable. It's not just a passing story on the news, but something you can see and smell, the aftermath in ashes, carried in wind for miles.

In August 2020, while the COVID-19 pandemic swept across the globe and the outdoors were one of our only respites, flames once again sparked across California. A series of fires were ignited by a dry lightning storm, engulfing three million acres of land. Many of these fires tore through rural land where crops—like pears, grapes, and otherwise—were growing. Many farmers lost years' worth of work, leaving us in fear of a food shortage.

It wasn't just in Northern California, either—the entire West Coast was on fire. There was no rain in sight and the blazes lasted for weeks. With smoke clogging the skies, people were advised to stay inside. We closed our windows, turned on air purifiers, and kept refreshing the Air Quality Index. On Wednesday, September 9, 2020, the Bay Area woke to a dark sky, the blue replaced by orange. The sun had been filtered out by smoke that had settled in low elevations. The sky remained orange for the entire day. It was apocalyptic.

But not all of us had the luxury to stay inside. This was during harvest season, and farmworkers needed to be in the fields to pick crops. They had no choice but to breathe the most polluted air the West Coast had ever experienced, as those fine smoke particles blanketed the earth—crops included—with ash. Farmers had to decide whether to sell or toss produce.

Fire season is supposed to end when the rainy season begins. But often the rain doesn't come, leaving us without the water needed to augment our snowpack or irrigate our crops, including pears. As the earth dries, so do these pear trees—and the farmers' income. For distilleries like St. George, this can mean trouble when sourcing ingredients. The trickle-down effect of climate change has implications at every level.

$\sim$

When I ask Lance and Dave about how fire and drought have affected the distillery, the mood grows solemn.

"You'll see some loss of harvest to the crop potential for pears, since they are where some of the fires have been," Dave says.

My question incites a different kind of response from Lance, who leans forward. "When it comes to those fires," he says, "our ability to harvest pears is probably the smallest concern that we could possibly have. There's so much other complete heartache and loss of life and property that takes place that I don't think we can worry too much about pears."

Dave nods in agreement. Both he and Jörg were personally affected by the 2017 fires in Santa Rosa, a small city in Sonoma County. After Jörg retired, he moved to Sonoma County, about sixty miles from where Dave's parents lived at the time. When the fires hit, Dave and Jörg were in Germany together on a distillery work trip.

"We were watching the fires go toward where Jörg lives, trying to get information in a different time zone," Dave says, his speech slowing from its usual quick cadence. "Finally, there was a wind shift that brought it back in the other direction. And then the fires were going toward where my parents lived. We were in shock by what was happening at home. So yeah, to Lance's point, we were less concerned about pears than about people."

His parents ended up evacuating, but they could see the flames creeping to the next tree line over from their home. Both their house and Jörg's were okay, but the emotional toll of the experience persists.

But before fire and drought ravaged the land in Northern California, pear trees were replaced by grape vines. In the early 2000s, Jörg, then still involved with daily operations, saw the writing on the wall; he realized that he needed to look elsewhere for brandy ingredients. He decided to source pears from Colorado. As the first person to open a modern craft distillery, Jörg spent a lot of time showing newcomers the ropes. He traveled from state to state, city to city, helping novice distillers find their voice. His ideas are woven into the DNA of many renowned distilleries, including Clear Creek in Oregon and Jack Rabbit Hill, maker of CapRock Brandy, its operation tucked in a delta in the shadow of the Rocky Mountains.

When Jörg was helping Lance Hanson set up Jack Rabbit Hill's operation in Hotchkiss, Colorado, he found a type of small pear that was similar to the ones that grow in Lake County and Mendocino County in California as well as the Black Forest back in Germany. "While he was out there, he found these amazing organic, dry-farmed pears that came from the western slopes of the Rockies," Lance tells me. When St. George started to lose their local supply, Jörg remembered this. Now they buy those pears from Colorado. They are smaller pears, which Lance describes as very aromatic. Their perfume, of course, shows up strongly in the bottle.

Sourcing pears from Colorado means compromising St. George's locally grown ethos, and it comes at an environmental cost. Once the pears are harvested, they have to make the 1,100-mile drive by freight truck to Alameda.

"From a carbon footprint perspective, it's awful, because we have to drive these pears all the way from Colorado to California when we used to have really good ones here," Lance says.

But the pears show up reliably each fall and the distillers don't have to worry about fire or drought or farmworkers harvesting them in hot, smoky conditions. This kind of trade-off is one that a lot of distillers have to make. Though Lance and Dave would much rather get pears from California, they are at the mercy of the regional agricultural system. If the orchards they've been sourcing from for years are disappearing to make way for another, more profitable crop, their choices are limited. They can either stop making it altogether, or keep buying locally, taking on a new varietal of pear that they don't like as much and that diverges from Jörg's original vision (which may very well cost them business), or they can look to other, farther-flung markets whose fruit they like and which practice organic farming. As small business owners, Lance and Dave opted for the latter.

"I think our pear brandy has never been better," Lance asserts.

Despite the scarcity of California-grown pears—those small, aromatic types that are full of flavor—Dave has kept trying to find local growers. Though he's currently buying from a single farm in California, Dave has been developing relationships with a few other growers and is finally able to secure a farm that can grow enough fruit to supply the needed quantity.

"We started to find some pears in California that are lovely and wonderful," Dave says. "We're finding those small pears that I feel strongly are exactly what Jörg was drawn to years ago, and 2019 is the first year that we're actually getting a good amount of California fruit back in probably over a decade. It's big."

While local fruit is a priority for Dave, so are sustainable growing practices. Though he and Lance prefer to use organic ingredients, each farm and each grower must consider a distinct set of circumstances when making the decision to use organic or conventional methods. This is especially true of pears—they grow on trees, which don't exactly lend themselves to crop rotation. It usually takes a tree three years to start sprouting fruit, but five to seven years to produce a full crop; trees usually live between fifteen and twenty-five years. So, while crop rotation isn't a part of Dave's conversations with the growers, their use of aids like pesticides is something they do discuss.

One of the key things that Dave looks for in a farm is transparency. The more information the farmers provide, the more open they are to answering questions, the more honest they are about their practices, the more Dave trusts them and their product, especially if he can't get to the farm on a regular basis. "Sometimes we will visit farms when we first start out a relationship, and then occasionally we might head back out there to check up on something, but realistically there's a side to it where you have a really good, open communication pathway with the growers who understand the farming practices," he says.

Dave's excitement about discussing St. George's growers becomes more palpable with each sentence. He talks faster, he leans in, one point turns into another.

"Frankly," he says, "I trust the communication stream. These farms are generally not large organizations. They are generally family-run and might have generations involved. You're usually talking to one of the main people—the brother or sister or uncle—and they're usually pretty straightforward."

Dave asks growers up front if they are certified organic. Some are, some aren't, and some aren't completely. Growers explain why they use organic practices in some parts of the farm, and why they don't in other areas with different varietals, and what their challenges are. Dave appreciates the details, especially when they concern systems that aren't governed by a vetted system such as certification. These open responses, the grower telling Dave something he might not necessarily want to hear, build trust. The growers understand their land, and how to manage it, and they don't hide behind greenwashing tactics.

There are barriers to organic farming. First, if a farmer has been using conventional methods and wants to switch to organic, there is a transition period. Without the help of fertilizers and other synthetic chemicals, the soil needs time to heal before it can produce the same yields. This means less product and less profit. Additionally, farmers face a learning curve as they begin to use organic practices, which can also lead to fewer crops. Off the field, there's a great deal of paperwork involved in the certification process, which a third party has to oversee. Finally, certification is very expensive and can be prohibitive. In fact, many small farms use organic farming methods but can't legally label their produce as organic because they can't afford to pay for the certification process. Dave considers these very real challenges when deciding whether he wants to work with a farmer.

"Do you monitor the weather where the crops grow?" I ask, trying to get a sense of how involved Dave is with the day-to-day operations of the farm, and how much he worries about the reliability of local pears in the face of climate change.

"They're generally the experts in their field," he says, glancing immediately at Lance, who smiles, the pun not lost on him. Satisfied with this reaction, Dave returns to his point. "They're the ones telling us what the harvest is going to look like, what the timing is, how things are happening. They're the people I trust to tell us what the yield looks like, what the potential flavors might be each year."

"We're distillers," Lance interrupts. "We've got other shit to do. I'm not going to kick a farmer's tractor tire just to make sure they are doing things right."

The trust goes both ways. St. George wants to maintain its relationships with these growers. Often, that comes down to money. Dave is committed to paying them a fair price for their pears. Some growers set their prices, but others will ask Dave to name his price. When this happens, Dave pitches the question back to them, asking what price they need. They'll usually tell him where they are with the crop that year, and where the market is.

"What they're telling us, basically, is that if a deal fell through, they still need to be able to move that fruit," he explains.

It's a delicate dance. Dave and the grower will go back and forth until Dave asks them what price they are going to get otherwise, without him setting the cost. He is straightforward with the grower, making it clear, often explicit, that he wants them to get a fair price that is good for them. He won't short them their livelihood to get a good deal.

"I know what I'm going to do with these pears," Dave says. "I know what our margins are going to look like. I'm not going to increase our margins and screw over the grower. I want to make sure that they stay alive and are able to continue producing great quality fruit next year and a decade from now."

Valuing a grower and their work is something that Dave is proud of, even if he sees this as self-serving. "I also really believe, selfishly, that it benefits us in the long run, because we're making sure that we can keep getting quality fruit so that we can make quality spirits. It's the only way for us to contribute to that system," he says.

The stakes are high for St. George when it comes to securing a steady supply of pears. This is especially true when it comes to maintaining the unique flavor profile of their brandy. Not every crop—of anything—is going to taste the same year after year. Pears, one year to the next, are going to have some variations. Pears grown in different parts of the country are going to taste different, too. Not every bottle is going to taste the same, either.

But distilleries need some level of consistency. Many, like St. George, have built their reputation—and customer base—on the flavor profiles of their products. This can be tricky to achieve if you approach spirits as agricultural products—dynamic things that were once alive, growing in the dirt.

To solve the issue of consistency, the thing that consumers have been conditioned to want, St. George makes more pear brandy than they bottle. They understand that not all pears are created equal. So, they crush, ferment, and distill eight to twelve batches of pears annually, each batch weighing between sixteen and twenty tons, which are then blended.

Blending is important because different crops of pears are likely to have different qualities. The goal is for the fruit to represent each harvest, and to taste like the one that came before, trying to stay within the framework that Jörg created years ago. In order to achieve consistency from season to season, they also blend in batches of pear distillate that they've set aside from previous years.

"The act of blending is what is really, really vital in hitting our flavor profile," Lance says. He looks for pears that have the right amount of sweet, honeyed aromatics to blend with others that have an almost metallic characteristic that makes the fruit beautiful and focused. "You need to make sure you have the right tools to blend properly. Those tools consist mostly of our noses and palates, and enough batches. You can use those batches to paint from different colors of pears and make sure you've got one that's well-rounded."

Lance, reading the writing on the wall, made connections with local bartenders. Though these bartenders were initially drawn to vodka—because it paid their bills—their curiosity about the pear brandy grew. Suddenly people were interested in cocktails again, drinking more than just vodka. It didn't happen overnight, but as cocktail culture seeped into dimly lit bars across the country, brandy slowly became a growing market for them.

"Eleven, twelve years ago, the idea that pear brandy was something that people were actually buying didn't exist as a growth opportunity," Dave tells me when I ask about the impact that cocktails have had on the distillery. "Now, we literally can't buy enough pears that are of the quality and type that we want to sell for the demand we have for brandy."

The trickle-down effect of climate change is here, too, in the blending phase. The fires that rage through Northern California each year leave their footprint on farm fields. The smoke that fills the air and the ash that rains down in its wake contaminate crops, altering their flavor. This is called smoke taint, and for beverage producers, it can compromise the quality of their product.

"Very sadly, this smoke taint makes relatively unusual fruit if you're working something down to the details like we do," Dave says.

Dave and Lance have considered distilling and bottling pears tainted with smoke. Lance sees the value of capturing the flavors of a compromised crop because it would be a snapshot of what the environmental conditions were like that year. It also brings income to farmers who can't sell that fruit to other people. Not only have their properties been impacted by the smoldering landscape, but they incur a financial burden, too. He wants to help them mitigate that, supporting the people who have supported his company.

"There's an important story to be told here," he says, removing his unlit cigar from his mouth. "This is what's going on in California. We're talking about how a bottle tells a story. This tells the story of a shit year when it comes to fire. If we bottle that, you can pick it up. You can sit there and smell it and taste it. It can bring you back to that place. It puts a pin in that story, and it raises awareness."

In fact, I tasted wine made with smoke-tainted grapes several years ago on a visit to the Central Coast, where a fire had plagued that year's harvest. The grapes absorbed the smoky air much like they did the microbes in the dirt. The ash clung to the grapes long after they were picked, crushed, and fermented. It haunted the bottle, each sip of the young wine a reminder of the conditions the grapes had grown in. I may not have been there for the fire, but the wine made me feel like I was, its presence strong in my wine glass.

Lance tells me he thinks they will distill and bottle a small batch of these pears in the future. At this point in California, as each year passes with both fire and drought, the effect of climate change is no longer just a looming threat. It's not a matter of *if* the next fire will hit, but when.

Water also returns to the story—except not in the field this time, but in the distillery.

When people talk about water in the booze industry, it's usually about the flavor it imparts, the ice, or the waste on the back end. Distillers talk about the water they use to dilute their spirits after it's been aged in a barrel, before it goes into a bottle to drink. Some people discuss water's mineral content and how it affects taste. But there is frustratingly little conversation about the sheer volume of usage, from start to finish, in the distillation process. I've yet to meet a distiller who can tell me how much water they actually use to make their products, expect maybe Scott Leopold.

And I've asked. Several times. Most recently, I sent a distiller friend a text asking if they could tell me how much water they use to make one batch of whiskey. "No," they texted back, "I can't. We share a pump with the business next door, so I don't really know." Forget about how much gets used to grow the grain, how much through the still, or how much is needed to clean everything up. Not just during a drought, but always.

Distilling also creates water pollution. Though people in the industry are tight-lipped about this topic, perhaps because it feels too big to tackle or too threatening to discuss, some researchers are exploring options for treating wastewater. One study, which was done by Yogita Kharayat and published in the *Journal of Integrative Environmental Sciences* in 2011, explores the consequences of the liquid waste and offers solutions for bioremediation. I doubt many distillers have read this study, save for, again, maybe Scott Leopold.

Water, and the lack of it, has long been on the minds of Lance and Dave. Like many distilleries, St. George uses municipal water, regulated by their local East Bay agency, and there aren't any mechanisms in place to reduce or conserve their usage. There's no cap on the amount they use; they just get charged for it.

Lance wants to change how they think about, and use, water. "In the interest of transparency, we need to get better at water conservation. We're not great at it," he admits, his brow furrowed. He wants the local water municipalities to put in systems that would help him, and everyone else, use less water, though he's realistic about how expensive this might be.

He also wants to install a water reclamation or cleaning system at the distillery. "Some sort of system," he says, "so that, when we're cleaning things off, we could reclaim it and use it again for cleanup." Something in the vein of what the Leopold brothers created for their family's distillery back in Colorado.

Water is his number-one environmental concern, both as a distiller and a Bay Area resident. "Distilling tends to be an embarrassingly water-intensive operation. When it comes down to it," he says, "in a real drought, all distilleries should stop functioning because we're taking a lot of water and using it for a while and then throwing it away"— even though, from a business perspective, it's just not realistic for them to shut down during dry years.

Dave feels the same way. "Water usage is 100 percent my main concern," he asserts. And because they share this concern, he and Lance are looking into a water conservation system that they can add to the distillery in the short run, over the next twelve months, before the fire season picks up again.

As with most things, cost is a roadblock. They start telling me about the ins and out of paying for a system like this, their feelings of guilt clinging to their words.

"It's really a matter of figuring out how to pay for this system, along with the other things we need to pay for," Lance says.

"As a small business, there's a constant push and pull of prioritizing infrastructure improvements," Dave says, mostly to me but also to Lance. "We need to make sure our employees get paid and there's money in the bank—at the same time that you're trying to add on something significant like water recycling."

"And," Lance adds, "those things help us feel better, but they don't sell more product."

Since St. George opened in 1982, the connection between raw ingredients and what ended up in the bottles continues to be at the heart of the distillery's work. Capturing the essence of a local pear is how they cultivated their identity, how they set themselves apart from other producers.

When I ask Lance to describe what the distillery smells like during harvest, he explains it like this: "Get a bottle of our pear brandy, pour some into a snifter, stick your nose in, and shut your eyes. That's exactly what it smells like at the distillery right now. And that's exactly the reason to make pear brandy, to convey a sense of thousands of pears sitting on the floor here."

Dave agrees. "Alcohol is a carrier pigeon for our real work," he says. "It's really just to capture the character of that pear, which lives and breathes on the skin." It may be romantic, but this ethos is evident in every bottle.

It's a tall order to maintain. No one lives in a vacuum, especially in the face of climate change. But Lance and Dave have risen to the occasion, and that has made their work a model for others who also need to adapt.

CHAPTER 6

Bartenders

One afternoon at the UC Berkeley campus, I'm sitting at my desk when I get a text from my husband, John. "Craig is working at Bar Agricole tonight. Want to meet me there for a drink?" I do, in fact. I don't get to Bar Agricole as much as I'd like—the place where the rubber meets the road. I can taste their house-style Old Fashioned as I type my response.

After work, I head for the SOMA (South of Market) neighborhood, a mix of sterile high-rise buildings and lofts housing tech companies. The bar is located on 11th Street, an area that lacks a steady stream of foot traffic and is notorious for car break-ins. As I approach the heavy glass front door, I immediately spot John sitting at the corner of the large L-shaped bar, already deep in conversation with bar manager Craig Lane.

Craig, who remains slender and brown-haired even as he approaches middle age, has tended bar here since Bar Agricole opened in 2010. It's been a decade of standing behind the concrete bar top, making drinks underneath large glass installations that hang from the skylights, and talking to customers who don't always get what this bar is all about.

Ten years ago, Thad Vogler and his business partners, including longtime San Francisco bartender Eric Johnson, brought Bar Agricole to life. Their philosophy was simple: carry only products that are made well, without artificial ingredients, highlight those quality ingredients, and showcase flavors that represent the regions where they are grown and created. Thad has referred to the latter as *regionalism*, a term that sometimes gets media attention but has been slow to catch on in the alcohol industry.

While their ethos is straightforward, running a bar this way has been anything but simple. Distributors who sell spirits to Bar Agricole don't always understand—or want to accept—that Thad and his team refuse to carry products with artificial additives, things that aren't a true representation of the products' raw ingredients. Customers don't always get it either, and are sometimes upset when they ask for a name-brand spirit only to be told that it's not available here.

"Thad has been really scrupulous about what spirits we carried," Craig says. "'What has caramel coloring in it? Let's get rid of it. What's the worst thing on our back bar right now? Why do we need it? Let's just get rid of it.' He is constantly asking those questions."

The staff doesn't always like it, either. Many young bartenders who haven't yet engaged with the idea of regionalism don't understand why Thad doesn't carry certain brands. Craig listens and tries to reason with them that more bottles on their shelves means more competition for the products that he and Thad really love. If customers are given less choice, they might be more inclined to drink things that have been carefully sourced.

Nevertheless, over the years Bar Agricole's reputation has grown as people have gotten more interested in cocktails, and, as a result, more adventurous, especially with the help of social media. Thad and Craig have taken to calling Bar Agricole a "farm bar," and they'll often use this as their hashtag on Instagram posts. For a time, the only beer they carried was a farmhouse ale made by Brasserie Lebbe, an independent farmer in France.

"The brewer is growing all of the barley himself, he feeds his goats the spent barley and hops after the beer is made, and his wife makes cheese from the goat milk," Craig explains. "He converts all of the goat waste into biofuel, which powers his farming equipment and beer operation. It drives the conversation about why we only have one beer on the menu. It's like, 'Well, if you can find us another beer maker who's growing all the barley themselves, we will gladly taste it and consider it for our list. But until that happens, there's no better representation of who we are at Bar Agricole."

Thad developed his commitment to regionalism over the years he spent working as a bartender, first at Yale University, where he went to college, and then at the San Francisco restaurant Farallon, where he worked intermittently between his travels across the globe, and finally at Slanted Door, where he began to put things in motion with the support of chef-owner Charles Phan. One example was when Thad called the beverage distribution company to come remove the soda guns. When the employee from the company arrived, he was unable to comprehend why Thad wanted to remove the guns and kept asking what was wrong with them and why they weren't working properly. Charles didn't question this, instead giving Thad the space to build one of San Francisco's most revered beverage programs.

Later, Thad consulted on several projects, including the now-closed Oakland-based Camino, a restaurant owned and operated by Russell Moore and Allison Hopelain, as well as chef Traci Des Jardin's Jardinière and Bay Area restaurateur Adriano Paganini's ground-breaking cocktail bar, Beretta. All of these projects allowed Thad to source spirits made by grower-producers (mostly from Europe, save for the occasional domestic brand like hometown heroes St. George Spirits) and only using organic citrus and syrups for the cocktails.

Craig first met Thad in the late 1990s at Farallon, which was nestled in the heart of the touristy Union Square neighborhood. "Thad spoke to my indie rock sensibility. He made me question why should we carry Grey Goose vodka. It's like Dave Matthews or something—they don't need any more of our money because they've already sold a million copies. If you really want to sell things that you believe in, get behind a band that speaks to you personally. I really responded to that," Craig remembers.

Craig signed on to open Bar Agricole because he wanted to push his creativity with cocktails. He had been running the bar at Farallon, long after Thad had left, and was experimenting with the same ideas, highlighting quality ingredients, but eventually felt like he was fighting an uphill battle. Customers were asking for brands by name, and it was tough to cultivate regulars because so many tourists flocked to Union Square.

One of the elements that Bar Agricole became known for was featuring long-forgotten classics on their menu. Craig and Thad and Eric would comb through classic cocktail books, mining them for underrated drinks that hadn't seen the light of day in recent memory. These drinks, like the Sleepy Head (a mixture of brandy, lime, mint, ginger, and sparkling wine), had yet to be played out.

"When we started, ten years ago, I was just trying to find a cool vintage cocktail that, we felt, hadn't been done over and over already," Craig recalls. "The Sleepy Head was one of my favorites for a long time."

Because there were few bells and whistles in Bar Agricole's cocktail program—no infused vermouth or fat-washed whiskies here—sourcing spirits and liqueurs were key to walking the walk. Thad thinks of spirits as archetypes—an original that has been imitated—and he has a habit of stripping down the layers of lingo slapped on a label by a marketing department so that the details of how it's made are evident. Over the years, he's indoctrinated his most loyal crew, including Craig.

"Thad is a little bit like a cult leader," Craig says. "He has such strong beliefs that you kind of either fall in line or you get out. I definitely got on board really early. I started to believe in regionalism—it spoke to me. I think it gave me a new lease on being a bartender, a career bartender. It didn't feel like a soulless job anymore."

If Thad is a cult leader, he's a fluid one, the kind who joins other cults himself. He's established incredibly strong relationships with several producers around the world. He has a long-standing relationship with Charles Neal, who imports French brandies, and he often takes core members of the Bar Agricole team on sourcing trips. He's been a supporter of Niesson Rhum Agricole because of their transparency. He's carried spirits from St. George for years, starting back in the early 2000s, when he managed the bar at Slanted Door and they were making vodka flavored naturally by infusing it with locally grown organic Buddha's hand fruit. He's been allied with Colorado's Leopold Bros. distillery since they began making spirits, prophesizing that master distiller Todd Leopold will save the soul of American whiskey. When Bar Agricole first opened, Leopold Bros. whiskey was the only one Thad carried.

Craig, who became a buyer for Bar Agricole, has learned to ask hard questions of each producer who wants to sell them spirits, in order to maintain the bar's identity and ethos. He wants to see how transparent a producer is about their process. How do they make their spirits? Where do the ingredients come from? Do they add anything artificial?

When I join Craig and John, leaning toward each from across the bar, I am greeted with a small pour of the rum they're discussing (from Craig) and a kiss (from John). Craig, like a seasoned bartender, leaves us for a few minutes so that we can settle in, and he checks on his other guests. A few other couples are scattered around the bar, and some small parties are cozied together in the tall concrete booths that line the dining room.

Craig returns a few minutes later, and John and I order a round of drinks. John asks for different cocktails each time we're here, letting his mood dictate his order. I'm a bit more predictable and always start with a rye Old Fashioned. The first time I had one here, I was immediately enamored by the nuanced flavors—a balance of spice from the rye, the bite from their house-made bitters, sweetness from the Leopold Bros. maraschino liqueur, body from the house-made gum syrup that they split with the maraschino, and brightness from the orange peel. It's one of my favorite cocktails, and better than most drinks I've ever enjoyed in bars all over the world; no one rivals Bar Agricole's Old Fashioned.

We clink our glasses together when the drinks arrive, happily letting the earlier part of the day melt away. As soon as Craig has a free moment, he comes back to check on us and the conversation quickly returns to the spirits he's most excited about. We talk about mezcal (he likes Mezonte), rum (there are a lot of regional variations with a complicated history), vermouth (if it's made with good ingredients, why infuse it with anything to cover the symphony of flavors?), and, of course, brandy. He shows us several bottles that I've never seen before, their labels scrawled with French words that I can't even pronounce.

Brandy represents Bar Agricole better than any other spirit, even though the bar is named after a style of rum. Through spirits importer Charles Neal, Craig and Thad are able to buy Calvados, Cognac, and Armagnac from farmers whom the pair refer to as grower-producers. They grow grapes and apples right there on their land. They keep animals whose milk they make into cheese. They use spent grain to feed them. They create a little ecosystem for themselves and make brandies that not only taste good but are true archetypes, their stories told in the liquid that swirls in each bottle.

"They are farmers first," Craig says. "They have a 200-year legacy of distilling in their family, and their connection to agriculture is pretty remarkable."

These stories are how Craig sells their cocktails. It's what he's doing when I arrive, and it's what he's doing when we talk spirits. It's never forced. Instead, it comes from a place of genuine passion, what he calls appreciation for the integrity of a spirit. He relies on this integrity to get his guests and other bartenders and servers excited about what they carry.

"There are so many different things involved in spirits that it can be baffling," he says, "but at the same time it appeals to me the same way as when I flip through record stacks for hours, looking for that jewel."

Many of these brandies vary from year to year because not every fruit crop nor every batch of alcohol tastes the same. This keeps Craig on his toes, especially when it comes to cocktail recipes. He and his team try different variations, carefully adjusting the proportions, or making ten versions with different vintages or brands of brandy.

It also keeps things interesting for guests. There's always something new to try, a different arrangement of ingredients, a new spirit on the back bar. It's the simplicity of what Bar Agricole does that can open new doors for people—a long-forgotten cocktail, a spirit made by a producer from a country they've never visited, a classic drink with house-made ingredients that dials up the flavor.

As I sip my rye Old Fashioned, I pause for a second to take the room in—the way the concrete bar feels underneath my fingertips, how the light bounces off the clear glass installation hanging overhead, the order of the small array of bottles lined up against the wall behind Craig. I'm glad for this moment because it will be my last there.

A few days earlier, Thad announced that they were closing that location, packing everything up, and moving to a new spot down the road where they'd have a bigger kitchen, more foot traffic, and the opportunity to focus more on the drink program. And a few weeks after that, as the COVID-19 pandemic took hold in the United States, they closed their doors indefinitely, with the future of the bar, and the industry, uncertain, just a few months shy of their tenth anniversary.

The spirit of Bar Agricole will live on, if not in its new home, then in the hearts and minds of the guests who came through to taste cocktails expertly made from the best products available, and the employees who spent their nights there, stirring and shaking, pouring and plating. And in me, who will never forget how good my last Old Fashioned tasted.

~

Responsible sourcing is one thing bars can do to shrink their environmental footprint. Another critical step is to reduce waste, a huge issue for the restaurant industry. On average, each year a single establishment creates 100,000 pounds of trash, while the entire industry produces upwards of 33 billion pounds of food waste. This is in the United States alone.

The waste stream is endless. Cardboard boxes and plastic wrappers used to protect bottles when they are shipped around the globe end up in the bin. So do the napkins that are deposited in front of guests and never used, the plastic straws that are slipped into tall Collins glasses, and the garnishes that are carefully cut, peeled, pickled, or dehydrated to make a drink look pretty. There are the desiccated rinds of fruit that's squeezed for juice, much of which is shipped from a different hemisphere because it's not in season locally. There are the berries that are macerated for syrup or muddled whole into drinks, the uneaten remainders tossed out with the melting ice. All those lime husks, mint sprigs, and egg shells add up.

The physical waste that gets carted to the curb each night isn't the only problem. There's also the energy that is used to grow the food and manufacture the straws, napkins, and bottles. According to FoodPrint, "Food production in the United States uses 15.7 percent of the total energy budget, 50 percent of the land, and 80 percent of all freshwater consumed," while "America wastes roughly 40 percent of its food." All that, and only about 5 percent of food is composted in the United States, making it the single largest component of the solid waste that ends up in municipal landfills. Food production is also the leading cause of freshwater pollution. To say this is a problem is an understatement.

Fortunately, some bartenders and owners are starting to reexamine practices that have long been ingrained in the industry. Usually one or two people begin to make small changes, which leads to more awareness, eventually collecting supporters in cities around the world, including Toronto, Los Angeles, London, New York, Chicago, Singapore, and San Francisco. I was introduced to three of them back in 2017 when all anyone wanted to talk about was the peril of straws after a video of a turtle with a striped plastic tube stuck up its nose went viral.

I met bar owner Ryan Chetiyawardana and Trash Collective founders Kelsey Ramage and Iain Griffiths at a beverage conference called Drink Chicago Style. The meeting was held in the hip Ace Hotel chain, located in a gentrifying part of town, where each session addressed big topics plaguing the industry, such as implicit bias and sustainability. I sat in a sun-soaked conference room, eager, despite the hard chairs pushed too closely together, to listen to four panelists discuss the steps they were taking in their respective bars to reduce waste. The moderator was as excited about banning plastic straws as anyone was that year, and ideas for eliminating them dominated her line of questions. But it was immediately obvious that the panelists—Ryan, Kelsey, Claire Sprouse of the Tin Roof Drink Community, and Mary Bartlett of LA's Ace Hotel—had already moved far beyond this limited topic.

Ryan, who is from the United Kingdom and is based in London, helms the lauded Mr. Lyan Group, a company that creates bars and consults on beverage programs (of which Iain was a co-founder). It all started in 2013, when many were boldly proclaiming that cocktail bars had seen every iteration possible and there was no more room to grow. Ryan took that assessment as a personal challenge. He began talking with Iain, whom he had met at Bramble in Edinburgh, Scotland, where they had both tended bar, about how they could build something that would move the industry forward.

"When people say, 'Oh, they don't know what they are doing,' why do we think *they* don't know what they're doing?" he asked. "It's crucial to challenge and check traditional standards."

Waste had been on Ryan's radar since he was a kid. He describes his parents as industrious Buddhists who scolded him when he left the lights on or the water running. He majored in biology in college and started to consider how everything is connected, including the carbon cycle. And then, when he started bartending at Bramble, he noticed how much they were throwing out at the end of each shift.

"We would throw out a bag of lemon husks or fresh juice we didn't use every night," he says. "I remember thinking, 'What the hell are we doing?' I started to question our whole setup, and it was those frustrations that influenced what White Lyan was as an idea."

The bar opened in 2013 on the only street in East London that, according to Iain, has yet to gentrify. (The bar's security camera footage has been subpoenaed three different times as evidence in crimes, including stabbings.) Being behind the door in a dodgy part of town gave the place an air of exclusivity. It was a pared-down bar that sat a handful of people, with a basement, complete with a stripper pole, designed for parties. They used no ice, no fresh produce, no branded bottles. Cocktails were batched, diluted with water, and then chilled. Not only did this cut down on water usage, but it allowed bartenders to get drinks out to guests faster.

Prior to opening White Lyan, Ryan had given a few talks on sustainability around the UK, but he was disheartened that no one from the industry seemed to care. Bartenders would dismiss his calls for implementing more environmentally friendly practices, such as ditching ice and fresh produce, because they seemed like the antithesis of hospitality. Colleagues told him that sustainability and luxury were mutually exclusive.

"I don't think luxury is a static thing," Ryan says. "It is like any other fashion—it changes. There's a myriad of influences that affect how it evolves. I don't think there was ever a point where luxury needed to be about opulence and waste. That is a self-imposed barrier that doesn't exist in the real world. We put those barriers in place in the industry."

For Ryan, luxury is something that feels special, purposeful, something that you want to return to, that isn't the run of the mill. He channeled this at White Lyan, where the consumers were unbothered by the decision not to use ice, fresh produce, or branded bottles. Instead, they were either curious about the bar's innovative techniques or simply didn't notice anything missing.

"The idea behind removing ice was that it chills and dilutes," Iain says. "It was innovation through control." Some, whom Iain calls "old school," weren't always into this idea, feeling like batched drinks killed the romance of watching a drink be made. "Bullshit," he says. "No one has even complained about getting their drinks faster."

Buying fresh produce for juice, syrups, and garnishes was expensive, and composting doesn't really exist on a commercial scale in London, so the bar used ingredients such as citric acid to balance flavor in drinks. Ryan also bought big batches of spirits delivered in twenty-liter vats directly from a few distilleries. (The details of this relationship aren't public due to a nondisclosure agreement.) Individual bottles became a thing of the past, reducing the bar's waste stream by about eighty-eight bottles per week.

The bar became a hot spot. Beyoncé threw her birthday party in the basement. Newspapers wrote features. Bartenders traveled across the globe to taste the drinks. It sparked fierce debates online about what a delicious drink is supposed to be, with famed bartenders weighing in on each side, including those who had never visited.

But not everything lasts forever. White Lyan closed in 2015. Iain chalks this up to the location or the fact that people like the buzz of a new place and then are onto the next one. "Sometimes it was packed and sometimes there were three people there," he says.

Ryan wanted to take White Lyan's concept and evolve it into something new. He wanted people to talk about drinks the same way they did food—where produce comes from, how it's sourced, how to cut waste, and how things come together when you look at all the links in the chain.

In 2014, Ryan, Iain, and their team opened Dandelyan, a bar in the upscale Mondrian London Hotel, overlooking the River Thames in the South Bank neighborhood. According to the Mr. Lyan's company website, the goal was to look at sustainability through the lens of the food system. The menu "explored the notion of industrialization, particularly as it pertains to botany."

I visited the bar one afternoon during the fall of 2017, while the sun was still hours away from setting. Everything about the place screamed luxury—the posh location, the tall ceilings, the glistening back bar, the sparkling glassware. John and I perched on overstuffed barstools and flipped through the detailed menu. It rotated drinks around seasonality and aimed to promote polyculture through the use of unusual ingredients such as mint, grapes, and hops. The menu's artwork was eye-catching, beautiful enough to frame.

John and I each ordered two rounds, slowly sipping our concoctions from the delicate glassware. As we traded tastes, John's eyes grew big as my eyebrows arched. We were both delighted at how nuanced these drinks were, at just how balanced the bartenders managed to make the flavors. We were impressed.

As John struck up a conversation with the dapper bartenders, their uniforms crisply starched, I scanned the shelves on the wall behind them, as I so often do. The sheen of my drink began to wan as I noticed brand-name spirits produced by big companies that source conventionally grown ingredients.

When I asked Ryan about this, he says he was frank with spirits companies about his interest in sustainability. In fact, he had given a talk at the annual bar conference Tales of the Cocktail in partnership with Bacardi—a company that mass-produces spirits.

"They were very willing to point out where they were failing," Ryan says. "They highlighted things that they needed to solve."

But why include them on Dandelyan's menu? Ryan explained, "It's about having a spectrum. You can't just have big and you can't just have small." He argued that small doesn't inherently mean better, and that while the big companies have flaws, they are making improvements.

It's a reasonable argument. Not all small distilleries produce sustainably. And there's value to pressuring large companies to change their practices. But it's also true that "green" aspirations can slip into greenwashing. A distiller will throw a solar panel on the roof and call the company solar-powered even though it uses gas-powered stills. Other brands say that they are zero-waste because they compost, even though they still generate lots of other trash. It can be hard for bar operators to wade through the marketing materials; it takes time that they don't always have.

I suspect that Ryan's decision to include brand-name spirits has more to do with time and money than with philosophy. Because the profit margin on cocktails is high, drinks are often how a bar that also serves food makes most of its money, so there is a lot of pressure to keep costs as low as possible because kitchen operations cut into the profits. Large producers cut deals to incentivize bar owners to carry their spirits. Choosing the cheaper, less-sustainable option is one of the many trade-offs that many businesses make to keep the lights on. In the UK, there is no three-tier system that governs the supply chain of alcohol between producer, distributor, and retailers like there is in the United States, so they were able to negotiate twelve-month contracts directly with brands.

~

When Kelsey Ramage began working at Dandelyan in 2015, she noticed the sheer amount of waste it was generating, which was typical for a bar its size. "It was such a behemoth of a bar," she says. "It was so big and we did so much volume and we were throwing out so much stuff."

Kelsey has an easy way about her—easy to smile, easy to nod, easy to engage. Maybe that comes from growing up in a small town—Salmon Arm, British Columbia, four hours inland from Vancouver on Canada's Pacific coast. She came from a household that recycled, learning at an early age how to separate plastic from paper. She felt that her small community had a better recycling program than bigger cities, where it wasn't as easy to monitor the program, educate consumers, and maintain the practice. She knew that ingredients came from gardens—they were all around her—and not just grocery stores.

At Dandelyan, she picked up what Ryan and Iain had started at White Lyan, driving the conversation about the wastefulness of bars forward. But there were no manuals to guide her, no forums for her to connect with like-minded people. Soon she and Iain (who were, for a time, a couple) were talking about the need for resources to provide bartenders and bar operators with a blueprint on how to reduce waste. They, like the rest of the industry, had been flying blind, trying to figure things out as they went along, and Dandelyan wasn't an open book.

"We realized there was a need to create a resource," Iain says. "If we were going to get bartenders to think about sustainability, Dandelyan wasn't going to be it."

Together, Kelsey and Iain began experimenting with recipes to get more use out of ingredients, like making a syrup from an avocado pit, trying single-handedly to divert a bit of trash from the waste stream. They started compiling these techniques on a blog in 2016, which they originally called *Trash Tiki* (in 2020 they changed the name to *Trash Collective*).

The *Trash Collective*'s tagline is "Drink Like You Give a Fuck." Kelsey and Iain wanted to infuse the blog with a punk attitude, channeling their musical taste while cultivating a distinct identity separate from the Lyan Company, which was all about luxury.

"Ryan is pretty highbrow," Kelsey says, "and we were going for lowbrow."

Iain describes the blog's tone—something they spent a lot of time thinking about—as aggressive, hoping to make the conversation around sustainability more fun and less stiff, but also careful to keep people from thinking that they were dumpster diving and that their drinks were unsafe to ingest. They curse excessively and disregard punctuation intentionally. It's full of DIY recipes intended to be made at home, without help from an army of trained culinary professionals.

What started as a simple blog quickly transformed into an educational platform to spread their ideas, recipes, and techniques. They traveled to different cities around the world and threw pop-up parties under the same moniker.

"Starting the Trash Collective was a bit selfish," Kelsey says. "We wanted to travel the world and see how different cultures were bartending and looking at ingredients. We wanted to answer the questions we were getting and put information out there."

Kelsey and Iain's work really shines when they work to educate bar teams about how to reduce waste. It's one thing to stand in the front of a classroom and preach to the audience about why being environ-mentally conscious matters, and it's another to guide them through it practically; they have to consider both when they're planning a pop-up. Most of the bars they collaborate with don't usually save ingredients to be reused. It takes some effort to get the ball rolling.

"We ask them to make a list of stuff they normally go through during prep, or even just send us their cocktail menu if they don't have time," she tells me. "Then we pull flavors from different things, like apple pulp that was left over from juicing."

They'd turn that into a syrup, concentrating the flavor that was left. "Or we'd do fermentation for *tepache* out of pineapple pulp," she adds. (Tepache is made from spent pineapples and brown sugar.) "It became very easy to do and we had quite a few different cocktails that were all derived from a classic drink in some way."

They'd work with the bartenders a couple of weeks in advance to prepare the ingredients, showing them how to get a second, third, or fourth life out of a fruit or a rind. Some of the bars adopted these meth-ods, weaving them into their regular repertoire. One bar in Hong Kong had been using coconut shells to hold drinks, but threw out the flesh. After working with Kelsey and Iain, they made Piña Coladas out of the leftover coconut meat.

Kelsey and Iain led a pop-up on the final night of the Drink Chicago Style conference, where I first encountered them, back in 2017. That evening, I ascended a darkened set of stairs in Chicago's Logan Square neighborhood. The East Room, which is now closed, was hosting the event, called "Wasteland Paradise," where people could try drinks made from "trash."

I found my way to the bar, where a handful of bartenders were run-ning back and forth, taking orders and serving cocktails. The menu was, fittingly, fun, and it didn't take itself too seriously. I shouted my order to the bartender, straining to be heard over the punk music that blared from the speakers. My drink came out fast. It was fruity and refreshing and did not come with a straw. I sipped on it happily as I joined my friends, all of us eager to recap the previous few days.

People danced around me, escaping into the night and making the most of our final few hours together in Chicago. We laughed and talked and drank, using the rest of our dwindling energy. Nothing remarkable happened—which, in itself, was remarkable. We sacrificed nothing—not one part of our experience—to spend the evening sipping things made out of "trash."

~

I caught up with Kelsey and Iain again in November 2019, where they were heading up a series of seminars at another conference, Portland Cocktail Week. It's modeled after a college program wherein participants apply and select majors, like Bartending & Hospitality, Bar Ownership, and Science & Technology. Since 2018, one of the majors has been Anti-Waste, led by Kelsey and Iain. This year, I was teaching a seminar in collaboration with Nicolas Torres, partner and bar operator of San Francisco's True Laurel. We were talking about spirit production, its connection to the food system, and why it's important for bartenders to ask questions about how products are made. There were a few other seminar leaders, including Brooke Toscana, who reduced waste at Pouring Ribbons, an iconic bar on the Lower East Side of Manhattan. There was also a cocktail competition for people to test their anti-waste skills. Kelsey and Iain would talk about recycling and the future of anti-waste.

"It's important to think and talk about ingredients, but we're realizing that reusing them doesn't have the greatest footprint," Kelsey says.

"It still requires a lot of sugar. You still have to buy other ingredients. I think we've shifted a little bit. When we first started, the Trash Collective was very much about ingredients, ingredients, ingredients. But now we're talking about spirits and purchasing and energy usage and recycling plastics. Seeing our carbon footprint versus reusing lime husks is so much greater. I don't think people realize that. This part of the movement is just starting."

This is true of Claire Sprouse, who is doing her part to educate the industry on environmental issues like plastic waste. On Tin Roof Drinks Community's website, she published a resource guide, complete with several articles about petroleum-based plastics, resin identification codes (RICs) that indicate what is recyclable, and carbon footprints, asserting that if a bar orders in bulk, it will result in fewer deliveries and less packaging. And yes, a lower carbon footprint.

When Kelsey took the stage at Portland Cocktail Week, standing in front of about forty bartenders, she launched into issues related to recycling and the waste stream, and how it's important to know the regulations of your local municipalities. There was no discussion of ingredients or making drinks from trash. In real time, we seemed to be watching her develop her understanding of a broken system where the same rules don't apply universally. She was teaching lessons about the environment, civics, voting, and consumer education.

I watched the faces of the bartenders who surrounded me. Many of them were learning about waste streams for the first time, and they were actually paying attention. Their eyebrows rose, their heads nodded, their jaws dropped. Their eyes widened when she spoke about asking brands not to send gimmicky trinkets made from rubber, or to reduce the excessive packaging on bottles. It only now occurred to many of them that they had the power to ask brands to reduce packaging.

The topic of water waste came up. In many parts of the country, it is common for bartenders to burn ice at the end of a shift—that is, to run hot water from a faucet over ice so that it will melt and disappear down the drain. Many people also keep water running continuously throughout an eight-hour shift so that they can easily clean their bar tools. Kelsey made the appeal for bartenders to stop wasting water, even though a few people in the audience couldn't comprehend doing things any other way. Then it was my turn to be shocked. As questions about wasteful habits rolled in, I discovered just how widespread and systematic these practices are around the world. It made me value the Trash Collective's work, and recognize how critical it is for industry leaders to examine their own habits.

~

In late 2019, Kelsey opened her first brick-and-mortar bar, called Supernova, in Toronto, where Iain was also a partner. It gave her a new perspective on operating a bar that uses sustainable practices. When I asked her about implementing some of the Trash Collective's anti-waste principles, she told me that it's always easier said than done, but sourcing and preserving ingredients are some of the bar's biggest efforts.

She buys her produce from 100KM Farms, a distributor that works with many organic and local farmers in the Toronto region, giving them a better route to market than what previously existed. It's not cheap, so she's trying to organize other bars around town to collaborate on orders. She has to go a bit farther for ginger, but works hard to create relationships that allow her to buy directly from the farms. She preserves short-season ingredients—peaches, plums, and raspberries—by adding sugar and vodka, turning them into a liqueur with a longer shelf life.

Lately, she's been looking beyond ingredients to spirits. She's been exploring how spirits are made and thinking carefully about what she wants lining the shelves at Supernova. She's taking a page out of Craig Lane's book—which, perhaps, will be the next iteration of the Trash Collective.

~

Back in London, Dandelyan closed its doors for good in early 2019 after four years. But Ryan continued to explore his ideas around sustainability, luxury, and blurring the lines between food and drink at Lyan Cub, which opened in the East London neighborhood of Hoxton in 2018. It's a partnership with Doug McMaster, who is behind Silo, a zero-waste-producing restaurant, also in London (and he's the author of *Silo: The Zero Waste Blueprint*). Ryan and Doug—who also is a friend of Kelsey and Iain's—had known each other for a long time, and they relished the opportunity to work together.

Cub was designed to challenge the divide between food and drink, something Ryan calls ridiculous and dangerous. The kitchen and the bar were equally important, working together to balance luxury and sustainability to create a fun atmosphere. The critics agreed, crowning it number twelve on London's "best restaurant" list; it was also the first venue to grace the cover of *Condé Nast Traveler*.

Things were going well until the COVID-19 pandemic hit. Sadly, Cub was among the virus's many causalities. Ryan made the official announcement in August 2020, saying, "We really believe in the values and systems it represented, and now more than ever, that seems to be a blueprint of how we want to change things. Empowerment, ownership of re-shaping value and luxury, human and ecological sustainability, and showing how food, drink, music, and community are such wonderfully warm tools to bring real happiness to people."

In Toronto, Supernova met the same fate. The pandemic made it impossible for Kelsey to keep paying rent on a space that wasn't bringing in adequate revenue. She made the announcement in July 2020, saying that she was heartbroken but hopeful for what is to come.

Though 2020 was cruel to the bar world, one thing is certain: the issue of sustainability is not going away. There is a lot of work to be done, and now models exist to improve sourcing and reduce waste. Each bar has to make tough decisions about how to go "green" and still stay afloat; as Ryan pointed out, it's a spectrum.

CHAPTER 7
Scale

As I'm writing this, it's summer 2020 and we're the midst of a global pandemic. Bars, at least the ones where we can sit inside, have been closed since March. I miss them. So I scroll through photos on my phone, taken long ago, or through my social media feeds, to remind me of a time when I could melt into a familiar stool after a long day.

In all these photos, mine or those taken by others, one thing strikes me: the bottles behind the bar look the same. Not the same layout or lighting or volume, but the same labels. Bars from Los Angeles to Boston to Miami to Chicago all buy the same brands, from the same companies. When people see those familiar labels at every bar they visit, it makes an impression. So they, too, buy those bottles at the liquor or grocery store. The scale at which Campari, Skyy, Evan Williams, and Bacardi are made is huge.

But the bottles you don't see with the same frequency, such as High Wire Distilling Co.'s Jimmy Red Corn Bourbon, the Leopold Bros.' Summer Gin, St. George Spirits' pear brandy, Mezonte's various mezcals, Equiano's Afro-Caribbean rum, aren't produced on the same scale. The giant companies and international conglomerates rule the industry, forcing the independently owned companies to grow and adapt, or perish.

What does this mean for the future of distilleries like High Wire and St. George? Are the business models of the Leopold Bros. and Mezonte financially sustainable? Will the ingredients that Equiano and Montanya use be available in the future? Can these brands make more and still protect the environment?

As these producers find success as independent businesses, they have to make decisions to ensure the longevity of their companies, from both an environmental and a financial perspective. Scale is an issue for all of them, one that each approaches differently.

The classic approach is scaling up—in other words, to make more. Making more comes with its own set of challenges, though, especially when it comes to the environment. Can distillers continue to use the same growers to produce ingredients without putting too much pressure on the land? Do they need to create relationships with new farms? How do they vet those new farms' growing practices? These are all issues that the owners of High Wire had to consider once they established themselves as eco-conscious producers.

"Jimmy Red Bourbon is the future of our distillery," Ann Marshall told me back in 2017 when I spoke with her and her husband, Scott Blackwell, at High Wire. Since then, they've gotten a lot of press and gained a strong fan base. Ann and Scott are very involved with the local food and beverage community, both in and beyond their home base in Charleston, South Carolina, and have given many talks about the environmental side of making Jimmy Red Bourbon, including one with me at the 2019 Tales of the Cocktail conference in New Orleans.

Things weren't perfect with their original way of operating. The amount of corn they were getting from the farms was inconsistent and they never knew what to expect. They were subject to the whims of the weather, making it hard to estimate how much bourbon they'd be able to produce. They wanted to continue to grow their business, to keep their production steady, so they decided to make some changes.

"We started to look at the supply chain. How much can we scale? How much do we want to manage? How much whiskey can we afford to sit on for years? There were a lot of considerations, and we had just gotten to a place where we felt comfortable," Scott told me when I caught up with him in August 2020.

One of those considerations involved the farmers who grow their Jimmy Red corn, farmers with whom they have close relationships. For instance, Jimmy Hagood, who showed us around Lavington Farms, has been a good partner, but he and his cousins—with whom he owns the property—don't irrigate their fields. In 2019, they experienced a drought and yielded five bushels an acre, which translates to just 280 pounds of corn. They only grew fourteen acres total and harvested less than 3,000 pounds. This wasn't enough for Ann and Scott to make as much whiskey as they wanted, even when added to the supply from other farms.

"To put that in perspective," Scott explains, "a GMO yellow dent no. 2 corn grown in Indiana will yield 250 bushels an acre. That's a big difference." He tells me that a small yield, like the one from Lavington, isn't good for farmers, either, because they can't make money.

On the other farms, Ann and Scott were seeing wild swings in yields. They might get fifty or sixty bushels an acre on one farm and ten on another. Like Lavington, all of the farms were close to the coast, where crops are at the mercy of hurricane season, the strong winds tearing through the fields and the floodwaters threatening to saturate the ground.

"We were getting pounded by tropical storms and hurricanes," Ann adds. "We were losing a lot of yield that way."

Ultimately, they needed farms with infrastructure that could help them scale up, and their current situation wasn't working for them or Lavington. After the bad growing season in 2019, Jimmy and his cousins took a vote.

"Jimmy can try to justify it to his cousins all day long," Scott says, "but if they are losing money, it's hard to validate."

Not every business relationship is meant to last forever. Though Ann and Scott and Jimmy had to move on, they still remain friends, and Ann and Scott still source cane juice from Lavington Farms. The couple also outgrew their relationship with Clemson University's research farm, though they still work with their crew in PeeDee.

Glenn Roberts of Anson Mills, who also still works with Ann and Scott, had been trying to get them to diversify their growing locations for a while, especially because South Carolina isn't the easiest of places to grow corn. So, after the 2019 harvest, they started working with four new partners, giving them a total of five contracted farms. These farms are still in South Carolina, but farther inland, off the coast, which mitigates the risk from hurricanes and pest invasions. Though the areas still get a fair amount of rain, it's cooler inland, so the growing conditions are better for corn.

"Moving inland has been critical to scaling up," Ann says.

With each passing season and another harvest under their belts, the couple's knowledge of farming deepens. Today, they are more familiar with what works and what doesn't, ensuring their Jimmy Red crops are successful, with stalks reaching tall into the sky and kernels turning a dark red. They learned that they needed irrigation and a regular watering cycle to make their corn happy and help it grow strong.

All five of the farms that they now work with irrigate their crops. Though manual watering increases the environmental footprint of the crop, the farmers try to offset this in other ways. The growers don't use chemicals, they don't till soil (so the top layer of earth isn't disrupted), and they support biodiversity by growing other crops such as lady peas, butter beans, and peanuts. These farms are growing food rather than commodity crops, so they don't live and die by the prices that are set by the market. This allows them to be stewards of the land and to keep growing for years to come.

"We have full-time farmers now that eat it, sleep it," Ann says. "This is their occupation. And they are blowing the roof off of it."

In 2020—a season that, according to Ann, everyone keeps calling a perfect year for corn because of optimal growing conditions in South Carolina—they were able to harvest ninety bushels per acre. Between all five of their farms, they laid down about 1,200 barrels of whiskey. That's a lot of bourbon, especially compared with the 225 that they made in years past.

"When we first started out, we were mainly using educational or hobbyist channels for grooming. That was critical," Ann says. "I can't stress how important that was because we couldn't have done this without Lavington and Clemson. They could do research at the same time they were growing, which informed a lot of our recommended farming practices now, like row spacing and watering and pest management. The baby steps we took were absolutely the only way we could have gotten to where we are now."

Though they've scaled up their growing operation, they're able to stay true to the ethos they adopted early on, like primarily using ingredients grown in South Carolina. They're distilling heirloom grains and have developed meaningful relationships with their farmers, who will often send photos of themselves in the fields next to a row of Jimmy Red, smiling next to the husks.

"We love Jimmy Red," Scott says. "It's a win-win that it tastes really good and it's of a place. There's pride in that."

~

As companies expand, so does their need for space. A facility that was once the foundation of the distillery, when ideas were still just experiments, may no longer be sufficient. So people need to move.

West Coast–based St. George Spirits did it. Founder Jörg Rupf started the distillery in a shack in Emeryville, California, in 1982 and grew the company into something that can barely be contained in a decommissioned airplane hangar. St. George's current location in Alameda, a few miles from Emeryville, is actually its third.

Jörg and master distiller Lance Winters successfully grew their business with vodkas during the Bay Area's cocktail revival of the early 2000s, and demand for their spirits grew nationally. They needed a bigger space to keep up with production.

"Our business plan called for eight hundred cases in the second year," Jörg recalls about the vodka boom in 2002. "We reached that goal in two months."

Two years later, in 2004, they moved from a smaller production facility in Alameda into a 65,000-square-foot airplane hangar on what was once the Alameda Naval Station. They had room for a bottling line; up to this point, they had been doing all their bottling by hand. There were offices and enough floor space for them to store barrels, bottles, stills, tanks, forklifts, a sugarcane press, and even a model shark from neighbors who built animatronic effects for movies. They had a place to ripen pears each fall and to blend each batch. They built a lab for research and development, a break room for the team, and a tasting room with giant windows that overlooked the distillery floor on one side and the San Francisco skyline on the other. The facility had a parking lot for visitors and a deck where people could gather outside.

The business blossomed in this space—and now, almost two decades later, it's just about bursting at the seams. They make five times as much pear brandy as they used to, not to mention their other products.

"When we first moved into this building, we went from 5,000 square feet to 65,000 square feet," Lance says. "We increased our amount of space by thirteen times. We never thought we'd be able to fill it up. Now we're trying to store a bunch of stuff so we can keep more barrels here. We're out of room."

When I ask if he'll move again, he says no. Instead, he has been thinking about moving parts of the production, like the whiskey or brandy, elsewhere. "But moving entirely? No, this place is too cool."

Others, like High Wire, that are just hitting their stride as distilleries, with their spirits finding their way onto more shelves, face similar dilemmas, no matter how cool their existing space. As their operation grew, Ann and Scott began to ask themselves some hard questions after the warts started to show in their original location. Could they increase production from five to seven days? Would this yield enough volume? Would another still fit? Did they want to stick with a pot still or pivot to a column to make each run more efficient? How would this change their process? Their flavor profile? Could they afford to hire more salespeople? The solution became clearer with each passing day.

Ultimately, in early 2020, they decided to let go of their space in the old Studebaker dealership and trade up for something new. They moved five blocks north from where they first started in 2012. In exchange, they gained almost 17,000 square feet. They had a custom pot still made for them because they liked the weight of the alcohol that comes off it.

"We didn't want to change our process or profile, because we like it," Scott says. The new still allowed them to keep the same flavor, but it holds four times the volume of their old one. They can fill almost six barrels a day now, close to thirty a week.

"That number really fits what we wanted to do," says Scott. "We can support ourselves. We can support a staff and salespeople and not have to sell a hundred thousand cases or something crazy like that. We don't have to break off a chunk of our business to sell it and wind up working for someone else."

In other words, they found their sweet spot. Just like Colorado's Leopold Bros., who carefully considered how large they wanted their operation to be when they moved from an industrial neighborhood in Denver to their facility in Jolie, just outside of the city limits, in 2014. Like Ann and Scott, brothers Todd and Scott Leopold want to remain independently owned, calling their own shots and making spirits that represent their values, philosophies, and regions. Just like Jörg and Lance.

But then, how do you stay relevant? When you no longer want to increase your production, how do you sustain your business model? All three distilleries have had to grapple with this question.

"The industry is in a crunchy place right now. We're seeing a whole lot of growth in small distilleries. There's a lot of noise. And the consumer tends to look for the shiny new thing. When you're a distillery that is over thirty-seven years old, you're not the shiny new thing," says Lance.

Head distiller Dave Smith and their team have worked continuously to maintain their segment of the market. Their production has increased and they're good at marketing, which has earned them loyal fans both locally and nationally. All of this helps keep them in the public eye. If they don't keep it up, their distributors won't pay as much attention.

"And then we start shrinking," Lance says. "It's a very competitive set up."

If their distributor loses faith, they run the risk of disappearing from stores and bar shelves, especially as the market gets more crowded. They need people to want their products, to ask for them by name, to pay attention.

To keep themselves competitive, St. George has put time and energy into developing new products, including their NOLA Coffee Liqueur, an homage to the city of New Orleans, where Dave met his wife, Julia. Or their Bruto Americano, a bitter *aperitivo* that some (like me) feel makes a better Negroni than Campari, not just because of flavor, but because, like all of St. George's products, it has no artificial additives. (Its color comes from cochineal beetles, not red dye no. 40.) When they launched Bruto in 2016, they threw a huge party at the distillery with freshly mixed cocktails, giant charcuterie plates, Bruto-flavored ice pops, T-shirts displaying the bottle label, and commemorative buttons with messages like "Bruto Americano: For a Bitter America" and "Bruto: Make Americano Great Again." It was a hopping party, the excitement thick in the air. Photos were posted on social media, which created buzz around the brand and kept distributors and consumers alike happy.

In fact, events like this one are another way that distilleries build a sustainable business model. This was a big consideration for Ann and Scott when they scaled up High Wire. They expanded their tasting room to four times the size of what it had been in the Studebaker dealership. They also now have a 3,000-square-foot event space, complete with a kitchen.

"We built this place with the future in mind," Scott says. "A big part of our marketing is agritourism because we want to showcase the agricultural side of what we do, especially since we see ourselves as food people first. We can have chefs do pop-ups here during harvest, host celebratory fundraisers, and do blending workshops."

It helps that South Carolina laws changed, allowing distilleries to mix drinks on-site. After the COVID-19 pandemic eases, they plan to rent the space out in order to diversify their sources of revenue.

Ann and Scott aren't the only ones thinking about diversifying their revenue stream. Some, like Todd and Scott Leopold, decided to sell not just alcohol but other things, such as malted barley, to local breweries and distillers to keep the cash flowing. The brothers built this source of income into their business model when they were planning a new malting facility. The new malthouse, which opened in 2020, is much larger than the malt floor they were previously operating next to their tasting room.

"It holds about 20,000 pounds of barley, and that's what goes on the floor at one time," Scott Leopold says. "It's almost twenty times bigger. It lowers the environmental footprint and produces a more flavorful grain."

The Leopolds don't need 20,000 pounds of barley for their own distillery, but if they can grow that much, why not sell it to others so they then don't have to buy from big industrial companies? Not only will selling malted barley to local distillers and brewers help sustain their business, but there will be a taste of Denver's terroir from the grain all over the region.

These are all ways for a distillery to stay independently owned, making their own decisions, charting their own future, and taking responsibility for their own success.

But what happens when you realize that you need outside investment to keep afloat? Or that you need more capital for improvements than a bank loan can provide? You might realize that selling shares of the company is your best option. If you're bought by a large company, what does this mean for scaling up production? Can you sell your distillery and still be environmentally conscious?

~

Maker's Mark makes bourbon. They have been making it in Kentucky since 1784, when Robert Samuels, a Scotch-Irish immigrant, set up shop and created his own recipe. Maker's became a commercial distillery in 1844, and in 1953, Samuels's great-great-great-grandson, Bill Samuels Sr., bought the Burks' Distillery in Loretto, renamed it, and added his own touch to the family's original recipe.

The distillery is nestled on a large campus in rural Loretto, sprawling across a thousand acres. It's dotted with dark-brown and red buildings—Maker's company colors—the newest of which is LEED-certified and built with energy efficiency in mind. It was built into a hill and one of the walls is limestone, designed to cool the building. The production facility houses several two-story working fermentation tanks made from cypress, which are official historic landmarks, and a stop on the Kentucky Bourbon Trail. Bill Sr.'s wife, Margie, came up with the brand's name, bottle design, and iconic red seal, earning her a spot in the Bourbon Hall of Fame as the first woman directly connected to a distillery (in large part because she was the first person to capitalize on "bourbon tourism" by starting the famed Bourbon Trail).

There's a lot of history here, evident in the white oak tree that has been rooted on the property for over one hundred years. But there's a mark of modernism, too, with green-engineered buildings, a ban on single-use plastic, and an employee whose job title is "Environmental Champion."

Many of these projects and staff positions wouldn't be possible without money from a larger parent company that can fund them. The brand has been sold several times, most recently in 2014 to Beam Suntory, one of the largest spirit conglomerates in the world, though the Samuels family retains creative control. This infusion of cash helped them expand the distillery, including the LEED-certified "green" building. The operation runs seven days a week, twenty-four hours a day.

But even with an operation of this size, the Samuelses make everything on-site. They don't contract industrial distilleries to make juice for them or outsource any of their production. They make their bourbon in batches of twenty-five barrels at a time and don't consider their whiskey to be mass-produced. In fact, they feel like they'll never be able to make enough to meet demand. Some top-level employees consider their bourbon to be as valuable as gold.

It's for these reasons—the billion-dollar company that owns them, the inability to scale up their production any further, the idea that their product is extremely valuable—that they've decided to invest in resource management. Their capital is going into conserving their natural resources—the very things that keep them alive, running, and profitable.

Jason Nally is the Starr Hill farm manager who oversaw sustainability operations and environmental education at Maker's Mark until summer 2020. He's been working there since 2017, after ten years with the Kentucky Department of Fish and Wildlife Resources as a private lands wildlife biologist. Jason leads a team of eleven who manage the property, and they see themselves as stewards of the land. He's also an educator for both the staff and the tour groups that come through to learn about Maker's environmental initiatives, and he works with teachers from local schools. (When he was promoted to farm manager, he passed the baton to Kim Harmon, who now heads the distillery's sustainability initiatives.)

"What we're doing means nothing if we can't educate people," Jason says.

Jason's investment in this land is personal. He grew up in Loretto, on his family's farm abutting the distillery, the property now owned by the company. He spent his childhood years roaming the woods and learning how the trails bend and the trees grow, how the water tastes, and which direction the winds blow. Now, most of his days are spent outside, trekking through the woods to monitor the progress of various projects, like establishing native grasses and building riparian buffers. He also works to improve Maker's infrastructure, eliminating single-use plastics and creating a recycling program for staff to bring their refuse from home because Loretto's municipal system lacks reach.

I meet Jason at the distillery on a cold day in February, a few weeks before the COVID-19 pandemic brought the world to its knees. There is a fresh coat of snow on the ground and although I'm immediately chilled to the bone, Jason looks at ease in this weather. He is in his early forties with sandy blond hair, smiling at me through his glasses, his gloved hands cupping a thermos of hot coffee. He welcomes me into his rugged golf cart, which has been outfitted for treks through the woods with its all-terrain tires and an enclosed heated cab that protects against the chilling wind. As we drive away from the dark-brown and red buildings and toward the woods, he points out different sites along the way that have both environmental and personal significance—an Abe Lincoln tour, of sorts. It's a bit disarming, the way he doesn't try to sell the brand by feeding me facts about sustainability. He's just telling me about the land. His conservation projects. How high the water has been in the lake lately.

This isn't what I expected. I'd been to Kentucky before, mostly to drive around the bucolic blue hills and drink bourbon at various distilleries. Kentucky is defined by bourbon, with its people wearing vests branded with distillery logos, billboards advertising competing companies that stretch endlessly along the highways, and the bars where it's normal to carry over fifty kinds of whiskey. There are seventy-three distilleries in Kentucky, each vying for your attention, each doing its best to differentiate from the next. I was expecting a canned PR spiel loaded with facts meant to convince me that Maker's is the best. Instead, I get Jason, who just wants to talk about the trees, the water, and the grass. He suspends my disbelief long enough for me to get past my preconceptions and actually listen to what he's telling me.

We stop a few times so he can show me the native grasses and riparian buffers that he's built, which are hard to see because of the snow that is blanketing the ground. We make a longer stop at one of the small lakes on the property, all fed by local groundwater. Water is integral to making any type of spirit, and this distillery has it in spades. They can access as much groundwater as they want, without paying the county or having water shipped in from some far-off place, like many distilleries do.

The water runs through limestone, which promotes good pond and lake health, imbuing the water with nutrients. It has low levels of iron and high levels of calcium and magnesium, which yeast loves. And the mix of minerals creates a Holy Grail of flavor. In fact, this water is one of Maker's biggest assets. They have more than 52 million gallons of spring-fed water available from their largest lake, which they use for both bottling and distilling, allowing them to produce over 2 million cases of bourbon in 2019.

"It's the not-so-secret ingredient in Maker's Mark," Jason tells me through his Kentucky drawl. "This has a lot to do with why the distillery is here in Loretto. It's not only the quantity of water that's vitally important, it's the quality of the water."

Jason started a project to look at how much water they use now, which is about 100,000 gallons a day, and how much they'll need in the future. He says they have a good idea of their short-term needs, but it gets more complicated when they're putting together their one-hundred-year initiatives. They consider hypothetical situations and consult hydrologists. They partnered with the Kentucky Water Resources Research Institute at the University of Kentucky and the United States Geological Survey, which has helped them with their water-sustainability models and provided an actual yellow submarine to give them a meter-by-meter reading of their lakes. This partnership resulted in the purchase of the farm next door, which was home to a lake that provided them access to an additional 32.1 million gallons of water, helping them protect the watershed and prepare for periods of prolonged drought.

The wind is whipping, so we head back to Jason's golf cart. By the time he parks it at our next stop, I can barely feel my toes. But Jason wastes no time exiting the cart and heads down a narrow path worn between the bare trees. I imagine how picturesque this must be in the spring and summer, the snow long gone, and try to trick the feeling back into my feet.

When we finally reach the place he wanted to show me, it's all worth it. Standing before us is a giant white oak tree, its limbs stretching impossibly high toward the clouds. Its thick trunk makes the younger trees growing around it look puny in its shadow.

"This is the Genome Tree," Jason says, switching into educator mode. "It's a white oak. I like it because this is one of the places I would sneak off to as a kid. This tree is important for us at Maker's because we're doing a complete mapping of its sequencing and genome."

"How old is this tree?" I ask as I crane my neck to look up at it.

"I think it's a couple of hundred years," Jason replies, "but it depends on who you ask. We're not going to core it with a borer to count the rings to find out, though—I can promise you that."

Natural resources are the backbone of the spirit industry, and it's not just the grain, soil, or malt that matters—it's the wood, too. White oak is valuable in Kentucky. In order for bourbon to officially be called (and labeled) "bourbon," it must be aged in new charred oak, the barrel freshly made specifically for this purpose. Ninety-five percent of the world's bourbon comes from Kentucky, and 1.7 million barrels were produced there in 2017, according to the Kentucky Distiller's Association. Production has increased by over 115 percent since 2012 as new distilleries have opened and more and more people have fallen in love with whiskey.

That's a lot of new barrels that are made from a lot of oak trees. In a 2019 interview for the *Whiskey Advocate* magazine, Jason had this to say: "The current supply of mature white oak is well established, but the concern lies in the next generation of white oak not being able to make it into the canopy. If we don't begin taking steps to solve the issue before it becomes truly problematic, the bourbon industry will eventually experience a barrel shortage, which will lead to an eventual decrease in production." Simply put, if forest conservation efforts are not made in Kentucky, there's going to be a lot less bourbon for us to drink in the future.

Here's the problem: younger white oak trees are not reaching maturity. They compete with maple and beech trees for sunlight, so they need to reach the forest canopy—the top layer of the forest habitat that provides shade and cover—to be successful in the ecosystem. Not to mention how the insects, invasive species, tree disease, and a warming planet can keep oak from reaching the canopy.

"They are also fire tolerant," Jason explains. "The absence of beneficial fires—what I call 'good fire'—is one of the reasons white oaks haven't been able to dominate the canopy."

He says that it's not necessarily about planting more white oak trees, but planting them responsibly and managing existing forests with an eye toward long-term sustainability initiatives, like controlled fires. Jason advocates for looking at trees like a crop, one that can benefit from the light touch of humans to help spur it along to ensure that the forest is sustainable for the next hundred years.

So here in front of us is this giant white oak tree, its trunk as thick as those of old-growth redwoods found in Northern California. This tree from Jason's childhood is now a precious resource, the object of a scientific study of its DNA. The distillery partnered with the cooperage that makes their barrels (the Independent Stave Company—ISC) and the University of Kentucky to research the tree's genetics and study threats from diseases.

Jason notes, "Disease is always a threat when you're looking at reforestation. This tree is the entire identity of the bourbon category." He explains that the majority of Maker's barrels come from a privately owned forest. The owners need to manage that forest sustainably for it to still be there a hundred years down the road. He concludes, "We all need to work together."

One of these collaborative efforts is the White Oak Initiative, a project that Maker's is undertaking with a few other distilleries and ISC—basically anyone who has skin in the game with white oak. According to their website, the goal of the project is to "ensure the long-term sustainability of America's white oak and the economic, social, and conversation benefits derived from white oak–dominated forests."

To learn more about the initiative, I'm scheduled for a tour of the Independent Stave Company, the world's largest barrel manufacturer. The company started in 1912 in the Missouri Ozarks before expanding their operation to Lebanon, Kentucky, in the 1980s. Between their two factories, they produce hundreds of thousands of barrels each year and employ over 1,500 people. They make barrels for distilleries, wineries, and anyone else who wants them. There are fewer than twenty barrel companies currently in operation, so many distilleries, including Maker's Mark and Leopold Bros., buy from them.

I leave Jason and his golf cart and am shuttled off to the ISC facilities in Lebanon, which feels a world away from the quiet beauty of the Maker's campus. With Mac trucks rolling by, we pull into the empty parking lot of a strip mall. We head toward a sign labeled "ISC Tours," and when we pull the doors open, there's a sterile room with a few clothing racks and a neat row of chairs arranged in front of a TV screen. A few people are there waiting for us, including our tour guide and Clint Evans, senior vice president of operations for ISC and Maker's point person for the White Oak Initiative. Both are dressed in fleece vests with Independent Stave Company logos emblazoned over their hearts.

The other person is Jane Bowie, Maker's "innovation expert." Her firm handshake matches her no-nonsense attitude and the passion for bourbon and wood that drives her work. Jane, now in her late thirties, has been with Maker's since 2007, working her way up the ranks. While much of the distillery is a well-oiled machine, Jane heads up experimental projects that sometimes lead to things like Maker's 46, a bourbon made by inserting French oak staves into a fully matured, cask-strength barrel. It's a different take on the classic bourbon that has given her the opportunity to stay relevant in a crowded market. It's also allowed her to dive deep into the world of wood, so she often finds herself in Lebanon to talk to the scientists at ISC.

We're shepherded through a tour of ISC—the only cooperage I've ever visited—where I'm given a headset and a pair of protective goggles. I can barely hear anything as our group is handed off from floor manager to floor manager, their rote speeches making it clear they have rattled off the barrel-making process countless times. Everywhere we go, we're surrounded by heat, fire, metal, and wood. The only women in the factory are Jane, our ISC tour guide, and myself. The only people of color work on various assembly lines, their pay dictated by the quality of the barrels they produce. When I ask about the ethics of this, I'm given vague answers that justify the company's pay practice and told there's a lot of proprietary information they can't discuss. But what is evident is the sheer number of barrels they produce—churning out one after another in a rapid, efficient succession.

After the tour, with my head swimming and my hair smelling like sawdust, we visit the lab. It's set up in a trailer next to the factory between stacks of oak. I'm offered a seat in the conference room where slides depicting molecules flash on the projector in front of me, the science of wood and flavor far beyond my understanding. The scientists can sense my confusion, so the conversation turns to their conservation efforts. This topic elicits a lot more transparency than ISC's working conditions.

I ask about the White Oak Initiative, which is Clint's territory. He's the chair of communication on their steering committee, so he's used to breaking things down in an understandable way for people like me, who aren't experts in forestry science. "We work with them on projects like reforestation and growing the next generation of white oak trees. Two years ago, we did a reestablishment of a forest."

They planted various species of trees, including white oak (oaks don't exist in a vacuum, after all), next to a stream that had been a pasture for years. Clint explains the details of the project, including getting his loggers certified in sustainable forestry practices. It's all about land management. The loggers are taught to properly "cruise timber," meaning they assess how much volume to take on a large scale without cutting down too much.

"We are trying to get more people interested in logging and forestry, because it's a dying industry," Clint says. "Many of our suppliers are fifty-five years old, or older. It's hard manual labor, and you're working in the hot and the cold. So we try to promote this the best we can, because if they disappear it'll be a real problem for our industry and proper forest management."

After my tours of ISC and the distillery, I'm convinced about the sincerity of the Maker's Mark environmental efforts. That doesn't mean their operation is 100 percent sustainable. For example, the corn they buy, even though it's from a family-owned farm, is not grown organically. The Petersons, who have been supplying the distillery with corn for years, would be willing to grow organic corn, but Maker's has never asked them to (although other companies have, and they do)—so it's a missed opportunity. If the conventionally grown corn doesn't meet the distillery's high standards, they send it back to the farm, putting the burden on the Petersons to figure out what to do with the unwanted kernels—sell or destroy them. At the end of the day, the yellow dent that Maker's uses is still a commodity crop. It's not all perfect, it's still a trade-off. Even with outside funding.

⁓

As craft distilling grows around the country, producers will increasingly face tough decisions about whether and how to scale up. But these are not the only questions. Sometimes the challenge is simply keeping what you have. Back in Charleston, Ann and Scott have been plotting the future of High Wire by looking to the past. They've realized that preserving the quality and flavor of their Jimmy Red Corn Bourbon means preserving seed security. They're working with grain geneticists and seed scientists, trying to find Jimmy Red's true type, a seed that will yield the same type of plant as the original.

"Corn has a mutt quality to it," Scott explains. "It cross-pollinates with other types of corn. During pollination, the polyp can drift or dormant genes can rise up. If you keep pulling that seed out of the field each season and throwing it in the ground the next, and the next, you'll end up taking a hard right turn. Before you know it, your corn is completely different than what you first planted—not just looks and color, but mainly flavor."

The idea is to protect Jimmy Red and the flavor they get from it. This quest has led them to library archives, history books, genetic labs, and to Mexico, where they are growing a few research crops to pinpoint its true type.

As Scott tells me about his conversation with an archaeologist in North Carolina, a grain geneticist who worked at Cornell University for thirty years, a germoplasm expert in Puerto Vallarta, and his theories about how Jimmy Red made its way from Mexico, where fifty-nine of the world's sixty-four varieties of corn originated, to South Carolina and into the hands of the enslaved people whose families preserved it, I realize that he is a de facto food historian. All in the name of flavor, all to protect the future of High Wire.

"We think of flavor as security," he says. "Being food people first, we know that what ingredients you start with matter for what ends up in the bottle. Agriculture tells a story."

This Is Not the End

When we think of the great environmental issues of our time, cocktails don't spring to mind. But the water, energy, pesticides, and other chemicals used to make a single drink do add up—not to mention the waste from packaging, bottles, garnishes, and more. The fact is that spirits are part of our food system, and that system has a tremendous environmental footprint. If you care about whether your eggs are free-range or your strawberries are organic, you should also care about where your drink came from.

Luckily, just as the good food movement has galvanized eaters, a good drink movement is beginning to take hold. It started in farm fields, distilleries, and bars like the ones described in this book. Though changing the unsustainable status quo will ultimately require action from producers, there's another group that has power—consumers: you and me.

The choices we make at the bar or liquor store are important. We can foster the success of small, sustainable businesses, like High Wire Distilling Co., St. George Spirits, and Montanya. And we can push large, well-known, and well-heeled beverage companies to "green" their practices. When we spend our money on a particular brand of spirits, it means we are investing in that company—and that means we should believe in the people and principles behind the label.

As consumers, we have the power to hold producers' feet to the fire. When distillers promote themselves as "green," we can ask them to pull back the curtain and show us how they are implementing sustainable practices. If they refuse to answer these calls, we have plenty of other producers to choose from—including Mezonte, Equiano, Leopold Bros., and yes, even Maker's Mark.

Competition is growing, along with the industry. In 2017, there were 1,500 craft distillers, up more than 20 percent from 2016. In 2020, the number had grown to 2,000—at least one in every state. And there's never been a better time for them to incorporate sustainability into their models. A growing body of nonprofits, advisory groups, and online tools are helping distillers and others in the industry adopt better practices.

Many of these organizations are focused on food, but some of them are now beginning to turn their attention to beverages. The more they do so, the less alcohol will be an afterthought in the larger conversation, and the gap will close between the kitchen and the bar.

The Sustainable Food Lab, for example, was founded in 2004 to support "green" and equitable practices throughout the food supply chain. The board leadership includes executives from large national companies, from the expected (Ben & Jerry's, Stonyfield Organic, and Oatly), to the more surprising (Target, Pepsico, and Starbucks). Alcohol is even represented by Anheuser-Busch InBev, the world's largest brewer. Their projects, which address topics such as organic grains and crop rotation, are just as relevant to spirits as they are to food.

As we've seen, a growing number of initiatives focus specifically on spirits. Drink organizations including the Trash Collective, Tales of the Cocktail, and Tin Roof Drinks Community are tackling sustainability, specifically for the bar community. Tin Roof Drinks, founded by Claire Sprouse and Chad Arnholt, national figures in the bar world (and winners of Tales of the Cocktail's first ever Sustainability Spirit Award in 2016), are keenly aware of what tools bartenders need in order to do their work. Their Cocktail Footprint Calculator, available for free on their website, is a spreadsheet that measures the environmental cost of each drink. Claire also spearheaded Outlook Good, a project that aims to take on climate justice in the beverage world. She put together a book of drink recipes called *Optimistic Cocktails* that the environmentally curious bartender can use as a guide or simply inspiration.

All of this work, coupled with the individual efforts of distillers and bar owners, is beginning to change the industry. By becoming a B Corp for accountability, working only with producers who value the environment, developing seed-security programs, or engineering a highly efficient distillation system, an innovative company can pave the way for others to not only follow in their footsteps, but to build on their models. There are many paths.

It is not always easy for distillers and bar owners struggling to keep the lights on to prioritize sustainability. It's simpler and cheaper to use the standard yellow dent corn or choose the labels that customers expect. But that's where consumers come in—we can change expectations. The more we ask of the spirits industry—transparency, responsibly sourced ingredients, less pollution and waste—the better off we'll be. And the better our drinks will taste.

Acknowledgments

This book wasn't written in a vacuum. Many people helped make it possible, and I'm grateful to each and every one of them.

First and foremost, thanks to everyone who spoke to me, taking the time to share their stories and the details of their work, and often hosting me along the way. The list is long, and includes Ann Marshall, Scott Blackwell, Glenn Roberts, Jimmy Hagood, Pedro Jiménez Gurría, Arturo Campos, Zule Arias, Scott Leopold, Ian Burrell, Karen Hoskin, Jörg Rupf, Lance Winters, Dave Smith, Craig Lane, Kelsey Rammage, Ryan Chetiyawardana, Iain Griffiths, Jason Nally, Jane Bowie, and their teams. This book, quite literally, wouldn't be possible without them.

Thanks to my agent, Matt Belford, who saw the potential for this book early on and championed it throughout the process. Thanks to my editor, Emily Turner, who helped shape this book into what it is and for always being patient when I needed to talk through a problem, and to the team at Island Press for giving this project such a great home.

Thanks to Rachel Khong and The Ruby, the communal workspace where much of this was written, to my writing group who read many drafts, especially Maggie Hoffman, Sarah Chamberlain, Emma Silvers, and the community that I found there. Thanks to Osayi Endolyn for her invaluable feedback and support, which helped this book see the light of day and has meant so much. Thanks to Sara Camp Milam and the Southern Foodways Alliance, who gave me time and space to think through the project in its early stages, through the writing workshop and on the *Gravy* podcast. Thanks to Stef Ferrari and Esther Mobley for allowing me to tell the stories that helped develop my voice. Thanks to my colleagues at the Oral History Center for the ways in which their work has helped me think about my own in new ways, especially on environmental topics.

Thanks to Carla Fresquez, Peggy Lee, Rita Bullwinkel, Nicola Parisi, Ellie Winters, Dave and Ada Freund, Vaughan Glidden, Mary Chiles, Cristina Kim, Jess Peterson, Adrian Hopkins, Wyatt Howard, Nicolas Torres and the others who have allowed me to talk about this book, my writing, and the ideas developed on these pages. I'm grateful to all of you.

Thanks to my family—Jessica, Juliana, Hudson, Lori, Cathy, Joe, Rachel, and Eames—and especially my parents, Jim and Suzanne, for their unwavering support and for being such fantastic cheerleaders.

And thanks, most of all, to John for the time, compassion, and support he's given me not just on this book, but every day. Thank you for always getting on planes with me, celebrating my wins like they are your own, and the unconditional love you've given me.

Bibliography

Chapter 1

Blackwell, Scott, and Ann Marshall. Interviews by Shanna Farrell, August 20, 2017, and August 5, 2020, audio.

Bliss, Jessica. "A Slave Taught Jack Daniel How to Make Whiskey. She's Made Telling His Story Her Life's Work." *The Tennessean*, February 23, 2018.

Bradford Watermelon Company. "Our Story." Accessed September 10, 2020. http://bradfordwatermelons.com/our-story/.

Brown, Elizabeth. "Stories of Trust: Jimmy Hagood." Low Country Land Trust (website). Accessed February 1, 2021. https://www.lowcountry landtrust.org/news/stories-of-trust-jimmy-hagood/.

Debczak, Michele. "This Super-Sweet Watermelon Has a Deadly History." *Mental Floss.* December 29, 2015. https://www.mentalfloss .com/article/71931/super-sweet-watermelon-has-deadly-history.

Distilled Spirits Council. "Distilled Spirits Council Forms New Environmental Sustainability Working Group." November 17, 2020. https:// www.distilledspirits.org/news/distilled-spirits-council-forms-new -environmental-sustainability-working-group/.

Distilled Spirits Council. "Sustainability in the Distilled Spirits Industry: Environmental Sustainability Best Practices." Accessed August 15, 2020. https://www.distilledspirits.org/sustainability/.

Hagood, Jimmy. Interview by Shanna Farrell, August 20, 2017, audio.

Johnson, Ayana Elizabeth and Katharine K. Wilkinson, eds. *All We Can Save*. New York: Penguin Random House, 2020.

McCrady, Allston. "The Return of Jimmy Red: An Heirloom Corn's Revival from Forgotten Field to Prized Bourbon." *The Local Palate*. Accessed August 1, 2020. https://thelocalpalate.com/articles/the-return -of-jimmy-red/.

Neimark, Jill, "Saving the Sweetest Watermelon the South Has Ever Known." *NPR's The Salt*, May 19, 2015. https://www.npr.org/sections /thesalt/2015/05/19/407949182/saving-the-sweetest-watermelon-the -south-has-ever-known.

Regenerative. "6 Problems with Monoculture Farming." Accessed September 10, 2020. https://regenerative.com/six-problems-mono culture-farming/.

Risen, Clay. "Can Liquor Have a Local Taste? They're Banking on It." *New York Times*, August 21, 2018.

———. "Jack Daniel's Embraces a Hidden Ingredient: Help from a Slave." *New York Times*, June 25, 2016.

———. "When Jack Daniel's Failed to Honor a Slave, an Author Rewrote History." *New York Times*, August 15, 2017.

Roberts, Glenn. Interview by Shanna Farrell, August 21, 2017, audio.

Saving Slave Houses. "Lavington Plantation." Accessed February 1, 2021. http://www.savingslavehouses.org/cw_2014jrh1029146/.

Tasting the South. "USC Expert David Shields on What's Missing from the Southern Food We Eat Now." *Post and Courier*, updated May 16, 2019. https://www.postandcourier.com/free-times/food/usc-expert -david-shields-on-what-s-missing-from-the-southern-food-we-eat-now/ article_49a6f7b5-c8e4-53b8-915b-0023cfa30798.html.

United States Department of Agriculture Marketing Service. "Yellow Dent Corn (Maize)." Accessed November 1, 2019. https://www.ams .usda.gov/book/yellow-corn.

Wilson, Victoria. "How the Growth of Monoculture Crop Is Destroying Our Planet and Still Leaving Us Hungry." *One Green Planet.* Accessed September 10, 2020. https://www.onegreenplanet.org/ani malsandnature/monoculture-crops-environment/.

Chapter 2

Beverage Industry Environmental Roundtable. "Climate Change & Scenarios." Accessed July 15, 2020. https://www.bieroundtable.com /work/climate-change-scenarios/.

Bowen, Sarah. *Divided Spirits: Tequila, Mezcal, and the Politics of Production.* Berkeley, CA: University of California Press, 2015.

Campos, Arturo. Interview by Shanna Farrell, September 7, 2019, audio.

Jiménez Gurría, Pedro. Interview by Shanna Farrell, September 7, 2019, audio.

———. *Viva Mezcal* (film). Directed by Pedro Jiménez Gurría, March 2016. https://www.youtube.com/watch?v=oqtXv2CAdGc.

Kharayat, Yogita. "Distillery Wastewater: Bioremediation Approaches." *Journal of Integrative Environmental Sciences* 9, no. 2 (June 2012): 69–91.

Leach, Kristyn, and Rebekah Moses. "The Future of Food." Interview by Julia Moskin of the *New York Times*, San Francisco Public Library, October 30, 2019.

Martineau, Chantal. *How the Gringos Stole Tequila: The Modern Age of Mexico's Most Traditional Spirit.* Chicago: Chicago Review Press, 2015.

Mezonte. "We Disseminate, Support, and Preserve the Traditional Spirits of Agave" (website). Accessed August 20, 2019. https://mez onte.com/en/mezonte-mezcal-y-otros-destilados-tradicionales-de -agave-english/.

Statista. "Consumption of Tequila in the United States from 2013 to 2019." Accessed November 5, 2020. https://www.statista.com /statistics/463120/us-consumption-of-tequila/.

Tetreault, Darcy, Cindy McCulligh, and Carlos Lucio. "Distilling Agro-extractivism: Agave and Tequila Production in Mexico." *Journal of Agrarian Change.* January 2021. https://onlinelibrary.wiley .com/doi/full/10.1111/joac.12402?af=R.

Wunderman, Ali. "3 Ways to Tell if a Booze Company Is Greenwashing." Liquor.com, February 20, 2020. https://www.liquor.com/green washing-liquor-brands-4796423.

Valenzuela-Zapata, Ana G., and Gary Paul Nabhan. *Tequila: A Natural and Cultural History.* Tucson, AZ: University of Arizona Press, 2004.

Chapter 3

Alter, Alexandra. "Yet Another 'Footprint' to Worry About: Water." *Wall Street Journal,* February 17, 2009. https://www.wsj.com/articles/SB 123483638138996305#:~:text=It%20takes%20roughly%2020%20 gallons,pair%20of%20Levi's%20stonewashed%20jeans.

Beer Advocate. "Leopold Bros. Brewery of Ann Arbor." Accessed November 1, 2020. https://www.beeradvocate.com/beer/profile /2979/?view=ratings&ba=macpapi.

Chefs Feed. "Scott Leopold." Accessed September 1, 2019. https:// www.chefsfeed.com/experts/4240-scott-leopold.

————. "Todd Leopold." Accessed September 1, 2019. https://www .chefsfeed.com/experts/4239-todd-leopold.

Chicago Bourbon. "Inside MGP: America's Most Mysterious Distillery." Accessed November 1, 2020. https://www.chicagobourbon.org /2018/10/23/inside-mgp-americas-most-mysterious-distillery/.

Fleming, Paddy. "How Much Water Does It Take to Manufacture a Gallon of Beer?" May 13, 2019. https://www.montana.edu/mmec /news/article.html?id=18735.

Graziano, Lisa. "Leopold Bros. Goes Old School in Quest to Make Better American Whiskey." *The Whiskey Wash*, March 3, 2015. https://thewhiskeywash.com/whiskey-styles/american-whiskey/leopold-bros-goes-old-school-in-quest-to-make-better-american-whiskey/.

Great Western Malting. "History of the Malting Process." Accessed September 1, 2019. https://www.greatwesternmalting.com/process/history-of-the-malting-process/.

Leopold Bros. "Our Story." Accessed September 1, 2019. https://www.leopoldbros.com/story.

MacLeod, L., and E. Evans. "Malting." *ScienceDirect* (website). Reference Module in Food Science. Accessed November 1, 2020. https://www.sciencedirect.com/topics/food-science/malting#:~:text=Malting%20is%20a%20controlled%20germination,and%20kilning%20to%20dry%20it.

Master Brewers Association of the Americas. "P-50 History of Barley Production in the USA." Accessed September 1, 2019. https://www.mbaa.com/meetings/archive/2011/Proceedings/pages/P-50.aspx#:~:text=Barley%20was%20introduced%20to%20North,in%20the%20early%2017th%20century.&text=Barley%20was%20first%20grown%20in,the%20colonists%20to%20produce%20beer.

McMahan, Dana. "1 Million Whiskey Barrels Come from This Distiller You've Probably Never Heard Of." *Courier Journal* (Louisville, KY), March 5, 2019. https://www.courier-journal.com/story/life/food/spirits/bourbon/2019/03/05/mgp-distillery-in-lawrenceburg-indiana-whiskey-maker-you-havent-heard/3033447002/.

Niemisto, Katrina. "Saving the Planet One Bottle at a Time: The Leopold Bros. Story." *Distiller Blog*. August 26, 2017. https://blog.distiller.com/leopold-bros-sustainability/.

Rasul, Nicole. "A Century after Prohibition Barley Makes Inroads in the Midwest." *Modern Farmer*, February 2, 2020. https://modernfarmer .com/2020/02/a-century-after-prohibition-barley-makes-inroads-in -the-midwest/.

Shapiro, Arielle. "Leopold Bros. Makes History with First Use of Malting Floor." *303 Magazine*, February 13, 2015. https://303magazine .com/2015/02/leopold-bros-makes-history-first-malting-floor-u-s/.

Stevens, Lindy. "Leopold Bros. Closes Doors." *Michigan Daily*, May 18, 2008. https://www.michigandaily.com/content/leopold-bros -closes-doors.

US Department of the Treasury, Alcohol and Tobacco Tax and Trade Bureau. "Limited Ingredients." Accessed November 1, 2020. https:// www.ttb.gov/scientific-services-division/limited-ingredients.

US Food and Drug Administration. "Compliance Policy Guide: CPG Sec 555.100 Alcohol; Use of Synthetic Alcohol in Foods." Accessed November 1, 2020. https://www.fda.gov/regulatory-information /search-fda-guidance-documents/cpg-sec-555100-alcohol-use-synthetic -alcohol-foods.

Weaver, John C. "United States Malting Barley Production." *Annals of the Association of American Geographers* 34, no. 2 (June 1944): 97–131.

Vogler, Thad. "Who Will Save the Soul of American Whiskey?" *PUNCH*, December 19, 2017. https://punchdrink.com/articles/who -will-save-american-whiskey-leopold-bros-rye/.

Chapter 4

Beverage Resources LLC. "FAQ." Accessed September 1, 2020. http:// www.beverageresources.com/index.php/sample-page/.

British Broadcasting Corporation. "Olaudah Equiano (c. 1745–1797)." Accessed August 12, 2020. http://www.bbc.co.uk/history/historic _figures/equiano_olaudah.shtml.

Burrell, Ian. Interview with Shanna Farrell, August 18, 2020.

Certified B Corporation. "Certification Requirements." Accessed September 20, 2020. https://bcorporation.net/certification/meet-the -requirements.

Curtis, Wayne. *And a Bottle of Rum: A History of the New World in Ten Cocktails.* Portland, OR: Broadway Books, 2007.

Eschman, Chad. "Put the Beetle Back in Your Negroni!" *VinePair* (website), November 20, 2016. https://vinepair.com/articles/now-you -can-put-bugs-in-your-negroni-just-like-the-old-days/.

Equiano. "Our Roots." Accessed August 12, 2020. https://equianorum .com/root/.

Green Groundswell. "Environmental Impacts of Sugar." Accessed on August 10, 2020. https://greengroundswell.com/environmental -impact-of-sugar/2019/07/22/.

Guardians of Rum. "About." Accessed August 12, 2020. https://guard iansofrum.org/about.

Hoskin, Karen. Interview by Shanna Farrell, September 28, 2020.

———. "In the Media." Accessed September 20, 2020. https://www .karenhoskin.com/in-the-media.

———. "The Rise of the B Corporation." *Distiller*, December 1, 2018. https://distilling.com/distillermagazine/the-rise-of-the-b-corporation/.

———. "Taking Sustainability Deep." *Distiller*, November 5, 2018. https://distilling.com/distillermagazine/taking-sustainability-deep/.

———. "Ten Ways I Freed My Distillery from Environmental Irresponsibility." *Distiller*, July 1, 2017. https://distilling.com/distiller magazine/ten-ways-i-freed-my-distillery-from-environmental-irrespon sibility-2/.

Ian Rum Burrell. "About." Accessed August 12, 2020. https://ianrum burrell.com/about.

Japhe, Brad. "Ian Burrell, Global Rum Ambassador and Equiano Rum Co-founder, Is 'Edu-taining' His Way to Racial Equity in Rum." *VinePair* (website). Accessed September 10, 2020. https://vinepair .com/articles/equiano-rum-ian-burrell/.

Lula-Westfield, LLC. "Production." Accessed September 20, 2020. http://www.luwest.com/default.php?content=production.

Minnick, Fred. "Rum Disruptor: Foursquare Calls Out Sweetened Rums and 'Absolute Poppycock' Perceptions." *Forbes*, March 29, 2019. https://www.forbes.com/sites/fredminnick/2019/03/29/rum -disrupter-foursquare-calls-out-sweetened-rums-and-absolute-poppy cock-perceptions/?sh=393986954c1b.

Montanya Distillers. "A Force for Good." Accessed September 20, 2020. https://www.montanyarum.com/a-force-for-good.

Omwoma, Solomon, Moses NyoTonglo Arowo, Joseph Lalah, and Karl-Werner Schramm. "Environmental Impacts of Sugarcane Pro- duction, Processing and Management: A Chemist's Perspective." *Environmental Research Journal* 8, no. 3 (October 14). https://www .researchgate.net/publication/272182420_Environmental_impacts_of _sugarcane_production_processing_and_management_A_chemist's _perspective.

Pardilla, Caroline. "Bars Toss Flor de Caña Rum Over Dire Worker Conditions." *Eater*, December 7, 2015. https://www.eater.com/drinks /2015/12/7/9838244/bars-boycott-flor-de-cana-rum-over-its-dire-work -conditions.

Pietrek, Matt. "How Foursquare Became the Pappy of Rum." *PUNCH*, March 21, 2018. https://punchdrink.com/articles/how-foursquare -rum-became-pappy-barbados-distillery/.

Potera, Carol. "Diet and Nutrition: The Artificial Food Dye Blues." *Environmental Health Perspectives* 118, no. 10 (October 2010). https://www.ncbi.nlm.nih.gov/pmc/articles/PMC2957945/.

Public Broadcasting System. "Olaudah Equiano." Accessed August 12, 2020. https://www.pbs.org/wgbh/aia/part1/1p276.html.

Radical Xchange. "Gimme Brown: Rum Talk" (live chat streamed online). June 30, 2020. https://www.radxc.com.

Rum Revelations. "Barbados Rum Identity—Richard Seale." November 17, 2019. https://www.rumrevelations.com/post/barbados-rum -identity-richard-seale.

———. "Geeking Out about Sugar Cane—Montanya's Karen Hoskin." October 14, 2019. https://www.rumrevelations.com/post/geeking -out-about-sugar-cane-montanya-s-karen-hoskin.

Schaffer, Ben. "Meet the Man Who's Shaking Up the Rum Industry." Liquor.com, December 14, 2017. https://www.liquor.com/articles /richard-seale-foursquare-distillery/.

Simonson, Robert. "The Secret to That Bright-Red Drink? Little Bugs." *New York Times*, November 26, 2018. https://www.nytimes .com/2018/11/26/dining/drinks/campari-color-aperitif-aperitivo.html.

Thorburn, P. J., J. S. Biggs, S. J. Attard, and J. Kemei. "Environmental Impacts of Irrigated Sugarcane Production: Nitrogen Lost through Runoff and Leaching." *Agriculture, Ecosystems & Environment* 114, no. 1 (November 2011): 1–11. https://www.sciencedirect.com/science /article/abs/pii/S0167880911002829.

US Department of the Treasury, Alcohol and Tobacco Tax and Trade Bureau. "Alcohol Beverage Authorities in the United States, Canada, and Puerto Rico." Accessed September 1, 2020. https://www .ttb.gov/wine/alcohol-beverage-control-boards.

US Food and Drug Administration. "Generally Recognized as Safe (GRAS)." Accessed September 1, 2020. https://www.fda.gov/food /food-ingredients-packaging/generally-recognized-safe-gras.

Wei, Clarissa. "The Silent Epidemic Behind Nicaragua's Rum." *Vice*, November 27, 2015. https://www.vice.com/en/article/qkxv7v/the -silent-epidemic-behind-nicaraguas-rum.

Wellness 360 Food. "List of Foods with Red Dye 40." Accessed September 1, 2020. https://iawpwellnesscoach.com/red-dye-40-foods-list/.

World Wildlife Foundation. "Sustainable Agriculture: Sugarcane." Accessed on August 10, 2020. https://www.worldwildlife.org/indus tries/sugarcane#:~:text=Sugar%20mills%20produce%20wastewater %2C%20emissions,leading%20to%20massive%20fish%20kills.

Chapter 5

Bland, Alastair. "California's Disappearing Apple Orchards." *Smithsonian*, November 1, 2011. https://www.smithsonianmag.com/travel /californias-disappearing-apple-orchards-125394178/.

California Department of Forestry and Fire Protection (Cal Fire). "2017 Incident Archive." Accessed on December 10, 2019. https:// www.fire.ca.gov/incidents/2017/.

Cornell, Maraya. "How Catastrophic Fires Have Raged through California." *National Geographic*, November 13, 2018. https://www .nationalgeographic.com/environment/article/how-california-fire -catastrophe-unfolded.

Frontline Wildfire Defense System. "When Is California Fire Season?" Accessed on December 10, 2019. https://www.frontlinewildfire.com /when-california-fire-season/#:~:text=Contrary%20to%20popular% 20belief%2C%20however,that%20blow%20across%20the%20state.

Kharayat, Yogita. "Distillery Wastewater: Bioremediation Approaches." *Journal of Integrative Environmental Sciences* 9, no. 2: 69–91. https://www.tandfonline.com/doi/full/10.1080/1943815X.2012.688056.

National Sustainable Agriculture Coalition. "Report Highlights Barriers and Opportunities for Farmers Interested in Organic." March 16, 2017. https://sustainableagriculture.net/blog/organic-transition-report-oregon-tilth/#:~:text=Obstacles%20that%20were%20primarily%20marked,such%20as%20seed%20and%20fertilizer.

Nicas, Jack, and Thomas Fuller. "Wildfire Becomes Deadliest in California History." *New York Times*, November 12, 2018. https://www.nytimes.com/2018/11/12/us/california-fires-camp-fire.html.

Nix, Joanna. "42 Dead, 8,400 Structures Burned, More than $1 Billion in Damage: the Devastating Toll of California's Wildfires." *Mother Jones*, October 25, 2017. https://www.motherjones.com/environment/2017/10/california-fires-damage-total/.

McKeever, Amy. "Foreboding Orange Skies Cast More than a Pall Over Northern California." *National Geographic*, September 10, 2020. https://www.nationalgeographic.com/science/article/foreboding-orange-skies-cast-more-than-pall-over-northern-california.

Rupf, Jörg. "A Distiller's Perspective on Contemporary Cocktail Culture." Interview by Shanna Farrell, The Oral History Center, 2016, audio, https://digicoll.lib.berkeley.edu/record/219015?ln=en.

Smith, Dave. Interview by Shanna Farrell, October 8, 2019.

St. George Spirits. "Story." Accessed on October 1, 2019. http://www.stgeorgespirits.com/story/.

Tucker, Jill, Dominic Fracassa, Dustin Gardiner, and Chase DiFeliciantonio. "Historic Northern California Fires Straining Resources as Newsom Calls for Help." *San Francisco Chronicle*, August 21, 2020. https://www.sfchronicle.com/california-wildfires/article/Growing-Northern-California-fires-straining-15505837.php.

Winters, Lance. Interview by Shanna Farrell, October 8, 2019.

Zhao, Duli, and Yang-Rui Li. "Climate Change and Sugarcane Production: Potential Impact and Mitigation Strategies." *International Journal of Agronomy* (October 2015). https://www.hindawi.com /journals/ija/2015/547386/.

Chapter 6

Bustamante, Lou. "With the Idea of Regionalism, Will More Bay Area Bars Champion Local Spirits?" *San Francisco Chronicle*, March 28, 2018. https://www.sfchronicle.com/wine/spirits/article/With-the -idea-of-regionalism-will-more-Bay-Area-12788203.php.

Chetiyawardana, Ryan. "Episode 28: Ryan Chetiyawardana aka Mr. Lyan on Rethinking Sustainability, Creativity and Growth, and Challenging the Status Quo." Interview by April Wachtel, Movers and Shakers Podcast, September 17, 2019, audio: https://www.movers sshakers.com/home/2019/9/6/ryan-chetiyawardana-aka-mr-lyan-on -challenging-the-status-quo-rethinking-sustainability-his-early-days-in -food-and-beverage-and-creativity-and-growth.

———. Interview by Shanna Farrell, August 5, 2020, audio.

Coffey, Helen. "Dandelyan: London Bar Crowned Best in the World Is About to Close Down." *Independent*, October 4, 2018. https://www .independent.co.uk/travel/news-and-advice/dandelyan-closed-london -bar-worlds-50-best-winner-mondrian-southbank-a8568571.html.

Do, Tiffany. "Tiki Takes the Food Waste Conversation Behind the Bar." *Food Republic*, August 21, 2017. https://www.foodrepublic.com /2017/08/21/trash-tiki-kill-sustainability/.

FoodPrint. "The Problem of Food Waste." Accessed on July 1, 2021. https://foodprint.org/issues/the-problem-of-food-waste/#:~:- text=The%20Problem%20of%20Food%20Waste,wasted%20in%20

the%20United%20States.&text=America%20wastes%20roughly%
2040%20percent,is%20perfectly%20edible%20and%20nutritious.

Griffiths, Iain. Interview by Shanna Farrell, November 1, 2019, audio.

Lane, Craig. Interview by Shanna Farrell, March 5, 2020, audio.

Lyan Cub. "Good Things to Eat and Drink." Accessed on July 10,
2020. https://www.lyancub.com/.

McMaster, Douglas. *Silo: The Zero Waste Blueprint.* E-Book. Brighton,
UK: Leaping Hare Press, October 2018. https://www.amazon.com
/Silo-Waste-Blueprint-Douglas-McMaster/dp/1782406131.

Mobley, Esther. "Thad Vogler to Relocate San Francisco's Bar Agricole,
Relaunch Obispo." *San Francisco Chronicle*, February 11, 2020.
https://www.sfchronicle.com/food/article/Thad-Vogler-to-relocate
-San-Francisco-s-Bar-15047612.php.

Mr Lyan. "About." Accessed on July 10, 2020. https://www.mrlyan.com/.

Ramage, Kelsey. Interview by Shanna Farrell, November 12, 2019, audio.

Raphael, Rina. "These Cocktails Are Garbage. Yum!" *New York Times*,
December 2, 2019. https://www.nytimes.com/2019/12/02/style/trash
-tiki-sustainable-drinking.html.

Simonson, Robert. "White Lyan in East London: A Bar beyond the
Status Quo." *New York Times*, September 8, 2014. https://www.ny
times.com/2014/09/10/dining/white-lyan-in-east-london-a-bar
-beyond-the-status-quo.html.

Stewart, Victoria. "Mr. Lyan on Why Sustainability Doesn't Mean Sac-
rifice." *Foodism*, September 18, 2017. https://foodism.co.uk/features
/interview-mr-lyan-cub-sustainability/.

Supernova Ballroom. "Supernova Ballroom Closing Announcement."
July 3, 2020. https://www.supernovaballroom.com/.

Tin Roof Drink Community. "Resources." Accessed on February 10,
2021. https://tinroofdrinkcommunity.com/resources/.

Trash Tiki. *Drink Like You Give a Fuck* (blog). Accessed on October 25, 2019. http://www.trashtikisucks.com/.

Vogler, Thad. *By the Smoke and the Smell: My Search for the Rare and Sublime on the Spirits Trail.* Emeryville, CA: Ten Speed Press, 2017.

———. Interview by Shanna Farrell, 2015, audio.

Walhout, Hannah. "His Bar Was Considered the Best in the World. Now, He's Starting from Scratch." *Travel and Leisure*, March 29, 2019. https://www.travelandleisure.com/food-drink/dandelyan-clos ing-lyaness-opening-best-bar-london.

Weiss, Zachary. "This Man Was Crowned the World's Best Bartender." *Observer*, July 27, 2015. https://observer.com/2015/07/this -man-was-crowned-the-worlds-best-bartender/.

Zimmerman, Liza B. "Is This the World's Greenest Bar?" Liquor.com. Accessed on July 10, 2020. https://www.liquor.com/articles/is-this -the-worlds-greenest-bar/.

Chapter 7

Bard, Jessica. "Marker's Mark to Install Solar Array, Research White Oak Trees." WDRB.com, November 8, 2019. https://www.wdrb.com /news/makers-mark-to-install-solar-array-research-white-oak-trees/ article_6fd98a02-0273-11ea-a794-cf064bc4c863.html.

Blackwell, Scott, and Ann Marshall. Interview by Shanna Farrell, August 5, 2020, audio.

Bowie, Jane. Interview by Shanna Farrell, February 7, 2020, audio.

Independent Stave Company. "Who We Are." Accessed February 15, 2020. https://www.independentstavecompany.com/.

Maker's Mark. "Meet the Maker's Mark Family." Accessed February 1, 2020. https://www.makersmark.com/ca/family.

Nally, Jason. Interview by Shanna Farrell, February 7, 2020, audio.

Stone, Sam. "Maker's Mark and the University of Kentucky Are Mapping the Genome of White Oak." *Whisky Advocate*, November 21, 2019. https://www.whiskyadvocate.com/makers-mark-white-oak -research/.

White Oak Initiative. "White Oak Initiative." Accessed February 15, 2020. https://www.whiteoakinitiative.org/.

About the Author

Shanna Farrell is an interviewer at UC Berkeley's Oral History Center, where she works on a wide variety of projects and specializes in drink's cultural and environmental history. She is the author of *Bay Area Cocktails* (History Press, 2017). Her writing has appeared in *Imbibe* magazine, *Life & Thyme*, *PUNCH*, and the *San Francisco Chronicle*. She holds master's degrees from both New York University and Columbia University.

Index